Dining Space Design

餐饮空间设计

黄玉枝 编

辽宁科学技术出版社
沈阳

序
Preface

随着台商在大陆各地的蔓枝散叶，也有越来越多的设计师由台湾跑到大陆来工作。这无疑是一种很自然的现象，因为作为服务业之一的设计师，既然台湾的老客户跑到大陆来了，自是难有理由不跟着过来试试水温的。但是，如果仅止于服务台商老客户，而得不到当地新客户的支持与鼓励，则企图在鱼龙混杂、竞争激烈的大陆占有一席之地，也不是件容易的事，更别说要开花结果了。

张秀贞设计师是转换相当顺利成功的一位。她固然是跟随着台商老客户的脚步而踏入上海，但在短短的三年间，却闯出了自己的一片天。她的客户群包括境内、境外、上海、北京、深圳等地的都有，大多是为其作品吸引而自动找上门来的。由此可见其作品受欢迎程度。

探讨其受到欢迎之道，大约可归纳成三点：一、是她只专心于做设计，而不涉及工程业务。她发现设计与工程在大陆犹如鱼与熊掌，很难兼得，因为跑工地的时间往往排挤掉了许多做设计的时间，于是她便舍弃工程而全身心全力地埋首于设计工作之中。二、是她乐意推心置腹地与客户为友，帮助客户慎选施工团队，并协助他们将工程尽可能地做到尽善尽美，虽然在这方面她并无利可图。三、是她能够化腐朽为神奇，将一些平凡的材料做出不平凡的效果来；也就是像一只变色龙似的，能在有限的预算内营造出一份空间的惊奇感来，这一点，对商业空间的设计而言尤其难能可贵。因为商业空间不同于天天要生活在其中的家居空间，它要有亮点，要能引起消费者的注意，要有话题可炒。一个商业空间如果能做到这点，其实已成功了一半，谨以此观，张秀贞商业空间作品中的那份惊奇感是很明显地不虞匮乏的。

现在，张设计师将其近三年来在大陆的作品汇为一集，呈现在社会大众面前，透过这本集子，读者将可以看出其室内设计师的功底，包括她对空间的巧妙安排，对材料的运用，对流行趋势的掌握，对生活的体验，对理想的坚持等等。而这也等于是一次勇敢的袒露，她以作品体现了自己，证明了自己。而我也相信，此举证明了她热爱工作以及她渴盼各界给予批评和教育的用心。（本文作者为台北《当代设计》杂志社总编辑）

—— 黄小石

With the spreading of the Taiwanese businessmen on the mainland, more and more designers came from Taiwan to the mainland for work, which is not so much as a very natural phenomenon. Serving as designers in one branch of the service trade, since their old customers in Taiwan have come to the mainland, it's only natural for them to follow their customers to have a try of the water temperature. However, if their service is only limited to the Taiwanese businessmen as their old customers and cannot find the support and encouragement from the local new clients, it is not an easy matter either for them to gain a foothold in the mainland where dragons and snakes are mixed together with fierce competitions, let alone blossoming and bearing of fruits in business.

Designer Zhang Xiuzhen is a quite successful designer with plain sailing in her transition of the track. Though she followed the footsteps of her old Taiwanese customers and arrived at Shanghai, yet she broke out her own path in business. Her group of customers includes the ones living inside the boundary of China as well as coming from the overseas, being located in Shanghai, Beijing, Shenzhen and some other places. Most of them took initiative to contact her because of the great attraction of her works, from which the warm response stirred up by her works may be clearly seen in part.

To explore her way how to win the warm response, the following three points may be summarized: First of all, she concentrated all her attention to the work of design with no involvement with the engineering business. She found that, on the mainland, the relationship between design and engineering is just like the one between fish and bear's foot, implying the difficulty for anyone to give consideration to both of the two matters, because it will consume a lot of time of design to make visits of the construction sites. Therefore, she gave up the engineering matter and devoted all her efforts to the work of design. Secondly, she is quite willing to make friends with the customers in a sincere way, offering assistance to the customers in the selection of the construction teams and helping them to make the projects to achieve perfection to the greatest possible degree, though she has nothing to gain in terms of economic benefits. Thirdly, she can turn the foul and rotten into the rare and ethereal so as to obtain the extraordinary effect from the ordinary materials, being just like a chameleon with the capability of building up a spatial sense of amazement within the limited budget. On this point, it is especially rare and commendable so far as the design of the commercial space is concerned, as the commercial space is different from the residential space in which one lives everyday. The commercial space requires bright spots for attracting the attention of consumers and topics for talking about. If a commercial space can achieve this, as a matter of fact, it has won half of the success. Viewing from this point, obviously, the sense of amazement in the commercial space works of Zhang Xiuzhen never lacks.

Now Designer Zhang has collected her works on the mainland in the recent three years into one volume for presentation to the general public. Through this volume, the readers may see her skill in the interior design, including her marvelous arrangement of space, application of materials, grasp of the popular trend of development, experience of life, insistence on ideal and so on. This also serves as a courageous exposure for her to demonstrate and prove herself through her works of design. I also believe that this act proves her love for the work and her intention of longing for criticism and advice from various circles of society. (The author of the preface serves as the editor in chief of Taipei Contemporary Design.)

—— By Huang Xiaoshi

前言
Preface

饮食文化是一个城市生活水平高低的指标之一。近年来，大城市的餐厅如雨后春笋开出，因应各类形态的菜色与不同的消费层，站在视觉与气氛营造最前线的空间设计，自此五花八门、别出心裁。餐饮空间较室内设计其他类别上更具商业挑战，设计会与人的情绪、感官、味觉甚至消费习惯产生化学反应，餐厅空间设计上的挑战，便是直接面对人的消费心理，是市场竞争的考验。

May Design 设计团队正是在这餐饮市场百花齐放的时刻来到上海，三年的创作与实战经验后，可说是经过设计检验与商业试炼的。因同台湾业主的配合机会而介入至大陆餐饮市场后，身为室内设计工作者的接触面更加宽广，除了室内设计的专业外，更加要懂"餐饮"行业。从资料的搜寻到市场敏锐度，都将成为设计空间时的必要考虑因素。

餐饮大环境的发展脉络

大陆正处于餐饮环境风起云涌的时代，随着经济的蓬勃发展，奢华风正盛行，餐厅的设计多少也受此影响。因此，May Design在处理餐厅设计案例时，基于尊重市场、尊重大环境的想法，是从调研餐饮市场做起，每个案例都是与业主共同互动成长。在台湾餐饮管理公司旗下的餐饮系统委托案例中，May Design借助其丰富、专业的市场经验，与业主取得共识将王品台塑牛排馆定位在商务市场客群，事后证明境内与境内商务客源非常平均地各占一半。

相较于境外业主，境内的业主对设计师的要求则是"拿出最好的"。May Design也就是因为先前的成功经验，而接连受到国内餐业主的青睐，比如豆捞坊的业主意在捕捉创造好奇心的投机市场；澜时尚料理的业主是以最好设计品质来要求设计师；北京悦堂火锅的业主以投资角色，会以策略规划来升级餐饮品项。

餐饮设计手法

最大视觉的占领

与业主共同找到客层定位是May Design的餐饮设计第一课。

首先将企业形象做为品牌塑造，在门店设计中以不同方式与最大面积地占领街廓视觉。以王品台塑牛排馆仙霞店为例，大幅的红色店招便矗立在两条路的转角建筑物上，在人潮川流不息的地段，很难被人们错失印象。西堤牛排馆苏州店的门面则内外环境合一。陶版屋的外观广告挑高处理后，显现出空间的魔力。泰平天国中，延续37米狭宽的半开放半透明门面，强烈的餐厅与酒吧混合的设计印象呼之欲出。北京悦堂火锅门面的星光大道，是以银黑亮丽光影铺开。

塑造精神图腾

善用图腾勾勒出餐厅的客层精神也是May Design常借用的手法。

西堤牛排馆的亮橘色系在桑巴风情的诠释下，呼唤着年轻白领的热情与活泼。苏州店中非洲菊图腾被大胆的用做铺面而成为精神标语。丰滑火锅苏州店中大自然的花卉是企业精神的代表，亦被巧妙的移用在火鹤的象征中，它从中国的水墨意象中抽离出来，形塑为灯笼高高挂起。泰平天国餐厅中的橙+灰是不张扬的泰式奶茶暗喻。澜时尚料理中蓝色的运用为接近苍穹的环境表现。陶版屋中紫红色丰满女人弹琴画，是视觉印象的符号。

平面规划非常灵活

由于业主在商业策略上的调整，一个非常灵活的平面与之对应，留给设计者日后调整设计的空间。

王品牛排馆在平面分割上，商业实验的支持下，坪效与设计的取舍下，设计者尝试将空间做不同的区分，以预留不同消费层与聚会时段的平面弹性。澜时尚料理中，电梯门厅是门面与过道的转换空间，在静与闹之间，是一种过滤不同平面气氛的灵活处理。

大量复制相同材质

数大便是美，在May Design中体现的是相同材质的复制。

王品牛排馆中大量运用金、银、米色推移西餐的优雅感，有大量复制相同之材质，如人造云石的天然质感与价位，是在模仿可控制成本前提下，对材质的复制。丰滑火锅中，量化材料、延续视觉的手法同样获得极好的效果。澜时尚料理中，四个圆形花朵形状的复制砌成玻璃杯架，取代了原先厚重的水泥柱。

成本控制

设计中的成本控制即为相似材料效果的取代。成本控制是大部分业主对设计者的要求，相对的也成了对设计的挑战。

王品牛排馆中，对艺术工艺的尝试也以效果取胜成本。玻璃夹金属丝、以铜丝线做双夹屏风，是初版设计的试验。西堤牛排馆中用到宣纸拼花效果，及以弹性布包裹柱子，同样达到对空间气氛活泼又稳重的要求。

功能 + 设计

在May Design的设计中，始终考虑对功能的同等重视。

在丰滑火锅中，一向为空间藏镜人身份的厨房出菜口，设计师将其以美观布置；男女厕共享的洗手台，则以椭圆符号做重迭设计。在各案例中运用最广泛的是不同空间的区隔手法，或以穿透珠帘、曲度皮编绳、锁链、玻璃，甚至灯光的明与暗，暗示空间的转换与过渡。高起呈波浪状的椅背设计，常常也是不同用餐空间的半区隔屏风。

有效的工程管理

May Design将工程管理视为与业主共建梦想的过程，从图面到结果，是实务经验与科技手法相加的过程。

引导客户对产品的定位、建立品牌定位，也是设计管理中很重要的一环。此时，在工程管理中最重要的是对时间的控制；全面操控合理的时间分配，将微妙牵动着设计的成本。设计与施工相为因果，因此在May Design的施工管理中，"设计必须符合施工可达成"会被列入管理的一环。

但May Design并没有因此而失去对设计的主导权，因为他们坚持设计者必须为施工者考虑施工成本被控制着必须达到合理性。他们有一套让设计空间更加自主的工作管理机制：严格的工程追踪与现场工程会议，有效的管理时间就是控制好成本，几个案子操作下来，便顺理成章的与业主建立了合作的信任感，这在设计行业中，是客户回流与人脉经营最重要的基础。

从设计端的梦想架构到空间实体的完成，其间工程施作被May Design看作是一个架设在建筑之下的"机械智慧"。事实上，在硬件装修中有三分之一的装饰符号是隐蔽工作，比如机电、空调的细节配置没规划好，会直接影响到设计的表现了。

结　语

餐饮空间设计是接受市场检验最直接、迅速的。从市场调查、品牌定位、设计概念的注入到施工完成，May Design在餐厅试营运期间可能存在的20%~30%冲突性，还将进行后续的设计版本的修正，这部分是因应市场运行的需求而生的。经过几次定版的最终定版设计，得到的现象是餐厅营业额上升，达成业主回收高的目的。对设计者而言，则是设计风格稳定的结果。May Design认为，经过严酷之业主、设计、施工考验后，真正成就设计的是餐饮市场的广大消费者。

Catering culture is one of the indices reflecting the high or low living standard of a city. In recent years, the restaurants have been opened in metropolises just like bamboo shoots after a spring rain, thus catering for various forms of cuisine and different layers of consumer layers.. The spatial design standing at the forefront of vision and atmospheric building thus becomes many and manifold and adopts an original approach. Compared with the other types of interior design, the catering space exhibits even greater commercial challenge. The design will produce chemical reaction on the human emotions, sensory organs, taste sense and even consumption habits; and the challenge posed to the design of the catering space is just to confront the consumption mentality of the people and therefore a test of the market competition.

May Design came to Shanghai just at the moment of blossoming of a hundred flowers in the catering market. After going through the dense creative activities and experiencing the actual battles, it may be said that our team has stood the test of design and tempering of commerce. With the chances of coloration with the Taiwanese proprietors entering into the catering market of the mainland, the surface of contact in the capacity of interior designers has become even broader. Apart from the specialty of interior design, it is even more necessary to gain understanding of the catering trade. Ranging from the search of data to the market sensitivity, they will all become the indispensable factors for consideration for the spatial design.

Vein of Development of the Grand Catering Environment

The mainland is positioned in an epoch of surging forward vigorously like rolling storms in terms of the catering environment. With the rapid development of economy, the wind of luxury is just prevailing; and the design of restaurants is more or less subject to this influence. Therefore, based on the idea of paying respect to both the market and the grand environment, May Design started its business from the designing cases of restaurants with every case interacting and growing with the proprietors. In the designing cases entrusted by the catering system under the flag of Taiwan Catering management Company, with the aid from the abundant and specialized market experience, May Design obtained consensus with the proprietor and positioned Taisu Wang Pin Beef Steak Restaurant as among the customer group of the commercial market. The ensuing facts prove that the sources of commercial customers both at home and abroad evenly cut halves.

Compared with the overseas proprietors, the inland proprietors set such demand on the designers as "Give us the best design." Also because of the previous successful experience, May Design found favor in the eyes of the inland proprietors of the catering trade. For instance, the proprietor of the Bean Curd House intended to catch up the speculative market full of creative curiosity; while the one of the Lan Fashion Restaurant required the designers as "Whatever design is wanted." However, the proprietor of Yuetang Chafing Dish Restaurant in Beijing acquired the knowledge of investment and knew how to make a plan with strategy for upgrading its catering items.

Catering Design Technique

Greatest Visual Occupation

Joint finding of the customer layer positioning with the proprietors is the first lesson for the catering design of May Design.

First of all, the enterprise image shall be used as the brand plastering. In the shop front design, the largest area shall be used to occupy the street vision by the different means. Take the Xianxia Shop of Taisu Wang Pin Beef Steak Restaurant as an example, a very large shop sign in red color was erected on a building at the corner of the junction of two streets. At such a road section with an endless stream of pedestrians passing by, it is difficult for them to miss the impression of the shop sign. The making of the shop front of the Suzhou Shop of Xiti Beef Steak Restaurant is the result of the merging of the internal and external environments. After the cantilevering treatment of the appearance advertisement of the Taobanwo Restaurant, its spatial charm is fully exhibited. For the Taiping Heavenly Kingdom Restaurant, its semi-open, semi-transparent shop front extending for a stretch of 37m in width, the designing impression with the strong mixture of a restaurant with a bar is vividly portrayed. The avenue of stars and lights serving as the shop front of Beijing Yuetang Chafing Dish Restaurant is a grand unfolding of lamp fixtures looking like an array of stars dotted in the sky.

Plastering of Spiritual Totem

To be good at using totem for depicting the spirit of customer layers of a restaurant is also a commonly used technique of May Design.

With the explanation of the Sangba flavor, the bright orange color system of Xiti Beef Steak Restaurant calls for the enthusiasm and liveliness of the young white collars. The totem of African chrysanthemum used in the Suzhou Shop as the shop front has become a spiritual slogan. The flowers and plants of Nature decorated in the Suzhou Shop of Fenghua Chafing Dish Restaurant serve as a representative of the enterprise spirit; and they have also been ingenuously transplanted into the symbol of fire crane, which was extracted from the Chinese traditional ink and wash image with the plastering of highly suspended red lanterns. The orange plus grey color system in the Taiping Heavenly Kingdom Restaurant is a metaphor of Thai type milky tea. The application of the blue color in the Lan Fashion Restaurant serves as an expression of the environment approaching the vault of heaven. The drawing depicting a well-shaped lady playing a music instrument in purple color is a symbol of visual image.

Very Flexible Planar Planning

Owing to the adjustment in the commercial strategy of an appropriator, a very flexible plane shall be correspondent to the adjustment, thus leaving space to the designers for ensuing adjustment in design.

In the planar division of the Wang Pin Beef Steak Restaurant, with the support of the commercial experiment as well as the acceptance or rejection of plateau efficiency and design, the designers attempted to make different divisions of the space so as to leave some planar elasticity to the consumer layers and gathering periods of time. In the Lan Fashion Restaurant, the elevator lobby serves as a transitional space between the shop front and the passage and between the quietness and noisiness, which is a flexible treatment for filtering the atmosphere of the different planes.

Reproduction of the Same Material Texture in Great Quantity

A large quantity will result in beauty; and in the May Design, it is reflected in the reproduction of the same material texture.

In the Wang Pin Beef Steak Restaurant, the golden, silvery and milky colors have been used in a large quantity for translating the sense of elegance of the Western cuisine. The abundant reproduction of the same material texture such as the natural textural sense and price level of man-made marbles is the reproduction of the material texture under the premise of simulating the controllable cost. In the case of the Fenghua Chafing Dish Restaurant, the technique of quantized materials and continued vision has obtained the extremely good result similarly. In the case of the Lan Fashion Restaurant, the glass cup shelf built in the shape of the reproduced four circular flowers take the place of originally thick and heavy cement pillars.

Cost Control

The cost control in design is just the substitution for the similar material effect. The cost control is a requirement of most proprietors on the designers and also constitutes a challenge to the design in a relative way.

In the Wang Pin Beef Steak Restaurant, the trial on the artistic technology also put the effect before the cost. The glass pieces with the built-in metal threads and the fabrication of the double-layered screens with the use of copper threads is an experiment for the initial design. In the Xiti Beef Steak Restaurant, the effect resulting from the use of rice paper for making of imitated flowers and the use of elastic cloth for wrapping the columns similarly meet the requirement on the spatial atmosphere of both liveliness and steadiness.

Function Plus Design

In the design made by May Design, consideration has been always given to the equal emphasis on the function.

In the case of the Fenghua Chafing Dish Restaurant, as for the dish outlet of the kitchen, the designer made the beautifying arrangement for it. As for the washing stand for the joint use by both men and women, the design was made with the repetition of elliptical signs. What is the most frequently used technique in various cases of design is the area-separating technique for the different spaces or use is made of the perforated pearl curtains, curved leather ropes, chains, glass pieces and even the brightness or darkness of lamp lights to denote the changeover and transition of the spaces. The design of the high chair backs in the wavy shape also serves as the semi-area-separating screens for the different catering spaces.

Efficient Engineering Management

May Design has regarded the engineering management as the process of building dreams together with the proprietors. From the drawings to the final result, it is also a process of the practical experience plus the scientific and technological technique.

Guidance offered to the customers in the product positioning and establishment of brand positioning is also an important link in the design management. At this time, the most important thing in the engineering management is the control of time. The overall control of the reasonable division of time will pull the cost of design in a delicate way. Design and construction serve as the cause and effect mutually. Therefore, in the construction management of May Design, the principle that design must conform to the possible achievement in construction will be listed as a link of the management.

However, May Design does not consequently lose the leading power of design, because they insist that the designer must be the constructors and the controlled construction cost must reach the reasonableness. They have a set of working management mechanism to enable the design space to be more autonomous, strict engineering tracking and site engineering meetings. The efficient management time is a good control of cost. With the completion of several projects, a sense of trust for cooperation will be naturally established with the proprietors. In the trade of design, this is the most important basis for the come-back of customers and the operation of inter-personal relationship.

From the construction of dreams at the design end to the completion of the spatial entity, the engineering construction between them is regarded by May Design as "mechanical wisdom" erected under the architecture. In fact, one third of the decorative symbols in the hardware fittings is the hidden work. For instance, the failure to make a good plan for the detailed arrangement of electro-mechanical devices or air-conditioning equipment will influence the expression of design in a direct way.

Conclusion

The design of catering spaces is the most direct and rapid in terms of accepting the market test. From the market investigation, brand positioning and infusion of the design concept to the completion of construction, 20~30% conflicts may exist for May Design during the period of trial business operation with the ensuing revision of the design edition made. This part of work occurs with the demand of the market operation. With the finalized design through several revisions, the resulting phenomena are the increase of the business volume of the restaurant in question and achievement of the purpose of the proprietor for the investment return. So far as the designer is concerned, it is the result of the stabilization of the design style. May Design holds the opinion that, after the severe test of the proprietor, design and construction, the genuine achievement for the design is the broad masses of consumers of the catering market.

目录
Contents

都会雅痞的菁英艺术

西堤 TAST 牛排

TASTY 是理性城市中热情雅痞，设计者将波普艺术的精神注入，图形象征是圆、菊腾、尚红橘色亮米及金属色。

几何科学运用从室内视觉引入情感知觉，率直开展出将方、圆、斜线痕迹交错于平立面上，处理元素由厚质感绿玻璃、平整镜面不锈钢、亮净深浅玻化砖，交错于层层相透玻璃之间。他、她的耀眼自信美由串串亮红闪烁红水晶来连续情境就是最最时尚元素。

菊腾图形、利用激光工业技巧切割红、黑、白、米色地砖，反差运用使单调框架激活出深浅律动。忽而翻覆于地坪规律、时而掠取吧台十字走廊的天空道，玻透垂直表现于彩玻、透明、明镜、墨镜等多种玻璃工业材料元素中。

TASTY is an enthusiastic Yuppie in a rational city. The designer infused the spirit of Pop art into it. The pictorial symbols are circles, chrysanthemum totem, orange color, bright cream color and metallic color.

The scientific application of geometry introduced the emotional senses from the interior vision with the straightforward unfolding of the traces of square, circles and oblique lines intercrossed on the planar or vertical surfaces. The treatment elements were composed of the green glass pieces with very thick textural sense, flat stainless steel with mirror surfaces and vitreous tiles with glossy hues in dark or pale colors being intercrossed among the transparent layers of glass pieces. His or her dazzling beauty of confidence was connected with the situations by means of bundles of brightly flashing red crystals, which became the most fashionable element in the decoration.

The chrysanthemum totem figures, the cutting of the red, black, white and cream-colored floor tiles with the use of the industrial technique of laser and the application of contrast made the monotone frames initiate a rhythmic beat. Sometimes, they rolled over the floor regularly; and sometimes, they captured the overhead passage along the corridor in front of the service counter. The glass transparency was expressed in a multiple of material elements of the glass industry such as colored glass, transparent glass, bright mirrors, ink glass and so on.

上海西堤新世界店

在新世界百货五楼，客源除了电梯的垂直输送外，并与位居百货公司与星级酒店双主入口地带形成一"T"形的动线，交点处为收银区。前后区块间夹带一长20米的电扶梯，此处安置为情人座区与45度斜面处理，使狭小空间看到不同宽度的明镜立面表现。在封闭空间中，以四分之一背对背的圆弧座位，协助空间的区隔与隐密性。

It is located on the fifth floor of the New World Department Store. As for the sources of customers, apart from the vertical conveying by means of the elevator, a T-shaped moving line was formed with the juncture of the dual entrance for the department store and a star-rated hotel with the cashier area located at the juncture. An electric moving of 20m in length was included between the front and back blocks. At this place the lovers' seats were arranged with the oblique treatment of 45 degrees so as to have a view of the vertical surfaces of the bright mirrors from the different widths of the narrow space. In this enclosed space, one quarter of the arc-shaped seats was arranged with back to back for assisting the spatial partition and keeping privacy.

主 设 计：张秀贞

参与人员：关中杰、赵立、吴习章

摄 影 师：刘圣辉

主要材料：米色烤漆、宣纸、金属环、丝线、方格架、明镜

坐落地点：黄浦区南京西路

面 积：925m²

完工时间：2005.04.28

协力厂商：上海高格建筑装潢

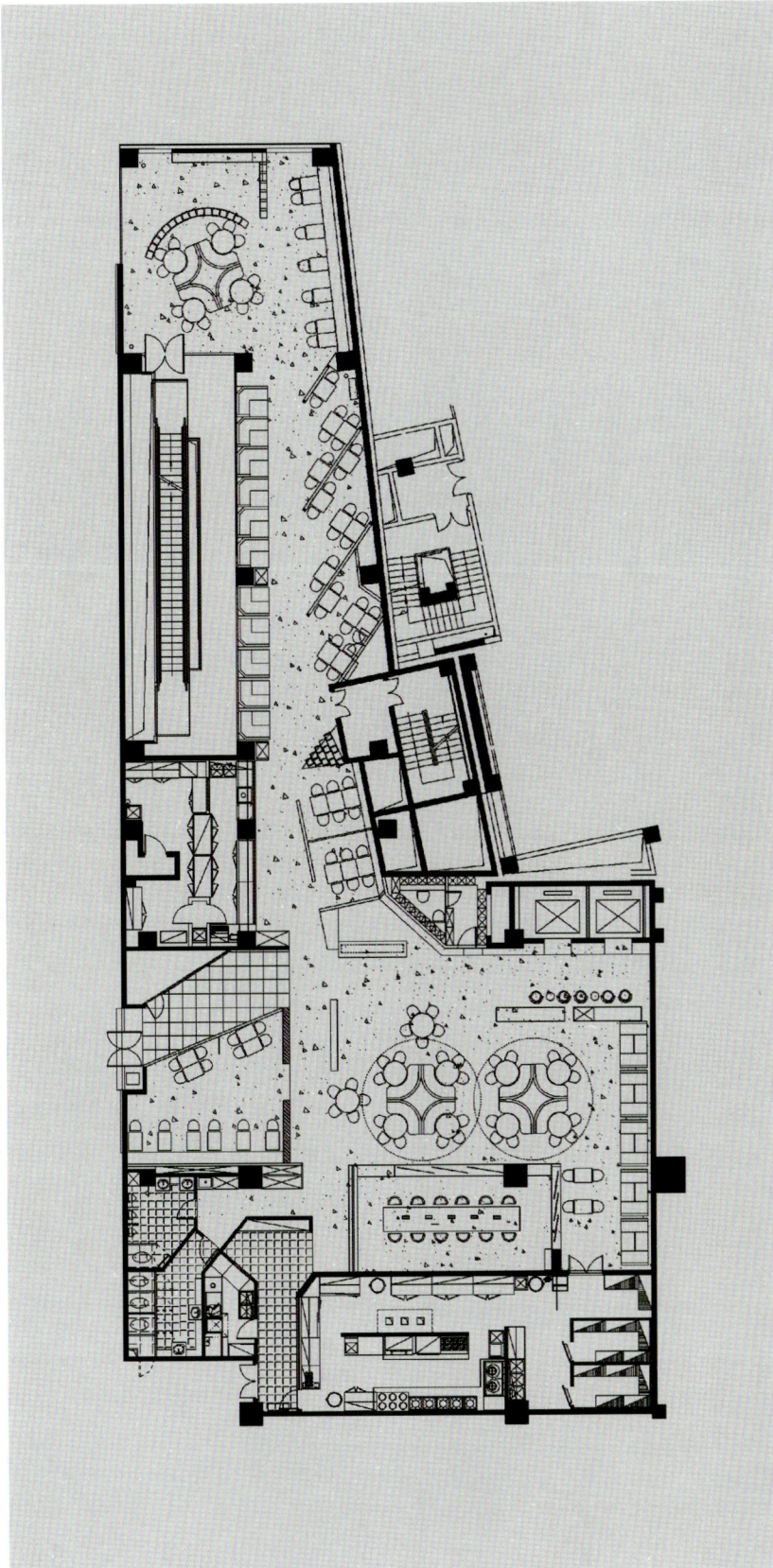

上海西堤天钥桥店

位于大街转角上，因是本系列的第一家店，外观上希望达到最大量体的视觉效应。在共享入口中做出自己的独立入口，并与位于四楼处高度的招牌形成对比。为化解长达约50米跨距动线的冗长感，在20米处做了一次客席的转折变化，斜拉出45度的隔屏。另外，玻璃光墙细腻处理了出菜通道的尴尬。

The restaurant is located at a street corner. As it was the first shop in the present series, in terms of appearance, it was hoped to obtain the shop focal vision effect with the largest dimensions. An independent entrance was built at the original entrance for joint usage and constituted a sharp contrast with the shop sign erected on the height of the fourth floor.

To dissolve the sense of tedious length with a span of moving line of 50m, a breaking change of customer seats was made at the length of 20m with the pulling of a partition screen with an angle of 45o. Besides, a glass wall offered a solution to the dilemma of the dish-serving passage in a very exquisite way.

主 设 计：张秀贞
参与人员：关中杰、赵立、吴习章
摄 影 师：刘圣辉
主要材料：米色烤漆、宣纸、金属环、丝线、方格架、明镜
坐落地点：徐汇区天钥桥路
面　　积：886m²
完工时间：2005.03.31
协力厂商：上海高格建筑装潢

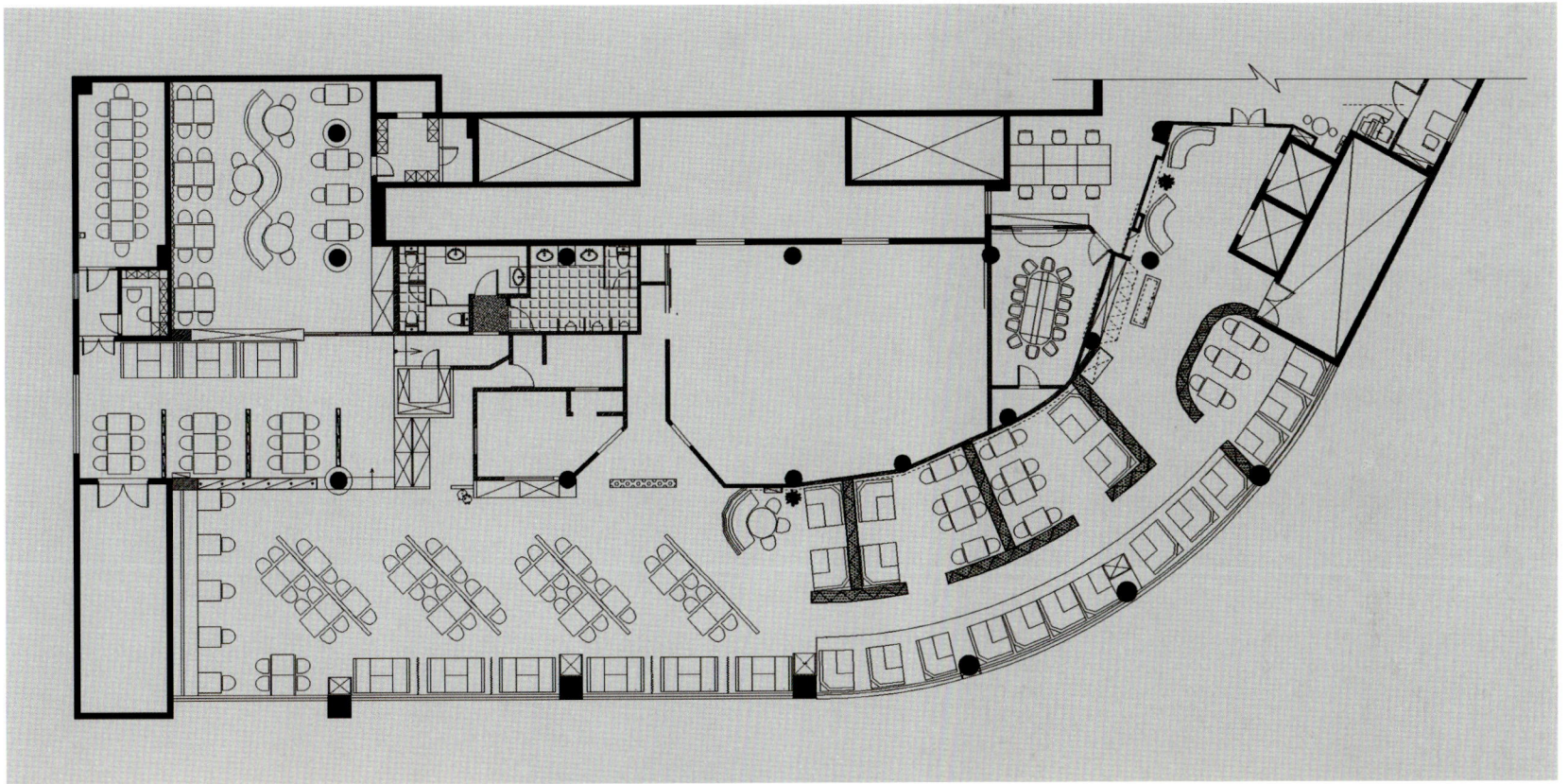

上海西堤淮海店

　　位于大楼三层，先封锁原靠淮海路的双向扶手梯的客源输送动线。采光面少，是因留外立面的广告招牌。为解决光线问题，再创三面宣纸投光的墙面，呈现户外有光景的质感。灯笼造型的柜台同样有协助光影表现效果。壁面橙色菱形软包有静音作用。四个柱列的连续符号与纵线的灯笼表现，成为柔化柱子在空间视觉障碍中最成功之例。

It is located on the third floor of a large building. First of all, the moving line for conveying the customers by the original bi-directional moving staircases adjoining the Huaihai Road was blocked. The surface of lighting was reduced because of the outer surface reserved for the erection of the advertising signboard. To offer a solution to the problem of lighting, the wall surfaces with reflection of light through the use of rice paper were created for providing a textural sense of outdoor light. The service counter in the shape of a lantern similarly had an expressional effect of assisting the light and shadow. The rhombus-shaped soft wrapping in orange color on the wall surface played the role of sound absorption. Four continuous symbols on the columns and the longitudinal lantern expression became the most successful example in softening the visual obstruction formed by the columns.

主　设　计：张秀贞
参与人员：关中杰、赵立、吴习章
摄　影　师：莫尚勤
主要材料：米色烤漆、宣纸、金属环、丝线、方格架、明镜
坐落地点：卢湾区淮海中路
面　　　积：1107m²
完工时间：2005.11.20
协力厂商：上海海银建筑装饰工程

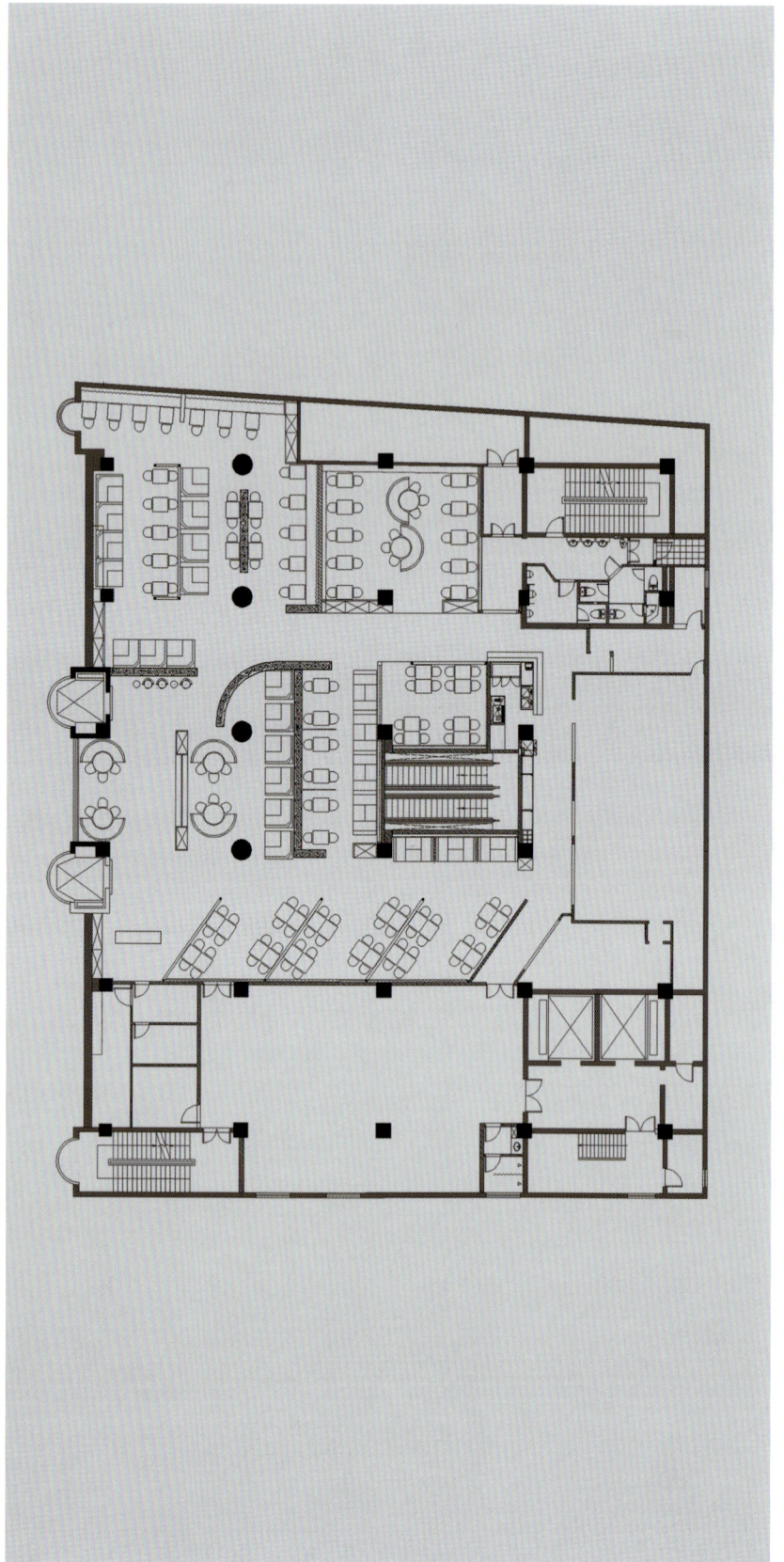

苏州西堤狮山店

　　本案靠近具有国际背景且现代气息较浓的苏州高新区。后退两米的入口设计，既在造型和空间层次上与"丰滑"形成了明显的区分，同时也避免了门厅在过道的局促感。L型转角处以烤漆玻璃配合LED灯凸显LOGO。左右两壁的波普风油画，以大胆色系演绎都会男女风情。自由曲度垂直而下的皮编屏风，后面的木皮编花，加以红色点缀，亦造成视觉上的现代印象。

The present project is in vicinity of the Suzhou High and New Technology Development Area with the international background and relatively strong modern flavor. The design of the entrance with the retreat of two meters formed an obvious difference from the Fenghua Restaurant in terms of profiling and spatial layers on one hand and avoids the sense of depression of the lobby set in the passage on the other. At the L-shaped turning corner, the vanished glass pieces were arranged to offer coordination to the LED lamps so as to made LOGO stand out markedly. The Pop style oil paintings on the right and left walls depicted the romance of men and women in bold colors. The leather-woven screens hanging directly downwards with free curves, a large wood skin woven flower in the back with the red-color decorations also resulted in a modern impression in vision.

主 设 计：张秀贞
参与人员：关中杰、赵立
摄 影 师：莫尚勤
主要材料：墨镜拼花、壁纸拼花、金属链、纱帘、烤漆图案
坐落地点：苏州市狮山路
面　　积：720m²
完工时间：2006.05.25
协力厂商：上海高格建筑装潢

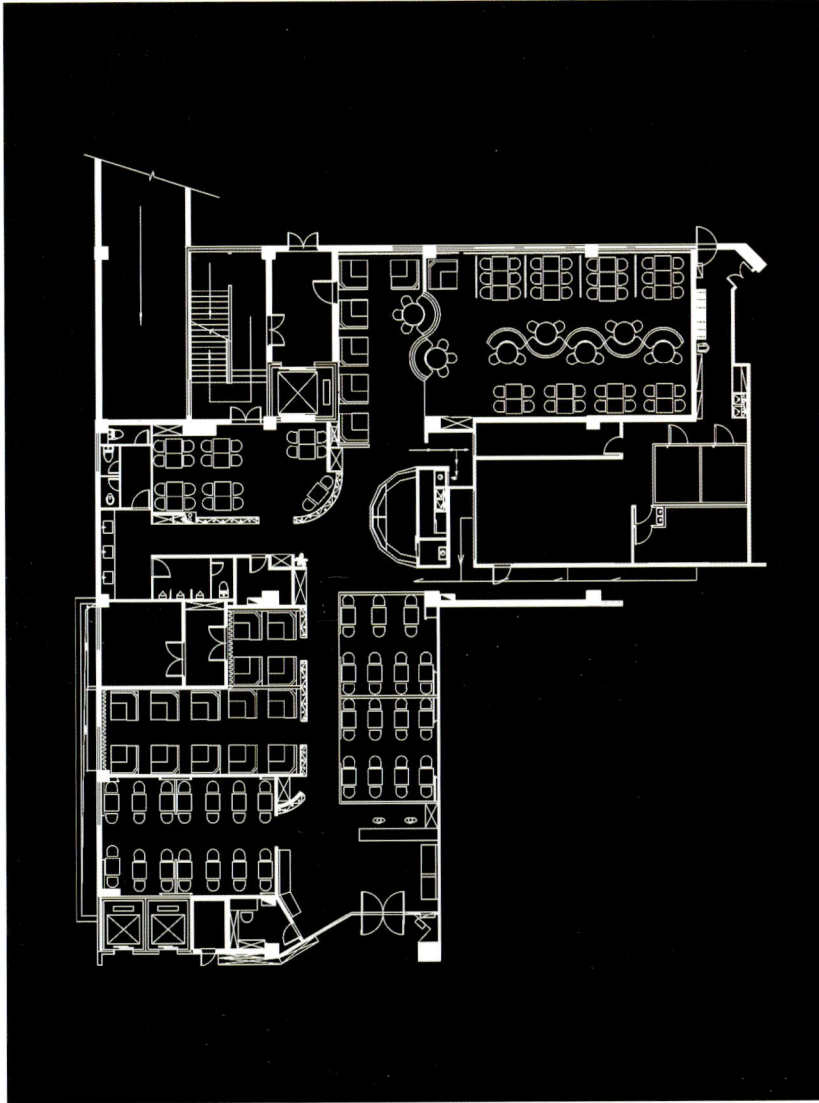

新古典主义的华丽诗人

王品台塑牛排

来自台湾的餐饮连锁体系 - 王品台塑牛排馆，是一家以展现经典优雅与上流品味的西餐厅。洋红金彩的招牌是品牌的标准色，在最突出的地点占领往来路人的视觉，一直是设计中最为重要的布局。

设计上为区隔出令人眼花缭乱、繁华的表像与唯美简洁的奢华之别，因此冷静处理出空间的流动与分割，才能在奢华的氛围中显现雅而不俗的风韵。运用流畅明亮的海浪艺术墙，张力无限衔接弧形大理石楼梯扶摇而上，金银双彩的金属丝玻璃屏风，不断以律动感呈现包厢空间的雅致。丰盛多样的客座变化，有皮质卡座、缎绒高背椅、尊贵头等舱椅，更有弧形紫红彩布卡座。在灯光下金链被照射的亮透，客席间再由布艺灯笼、软帘作为私密性及静音性的区隔，在材料上，更加强吸音元素，达到宁静用餐的优质环境。

闪亮的铝丝灯带、彩绘玻璃餐盘也成了艺术展现的材料，洗手间文化是现代餐饮空间最注重的一环，特以沙金玻璃墙及复式台面布置，贵气十足的金框门套，再现经典的红金色交错出新古典的浪漫。

Taisu Wang Pin Beef Steak Restaurant, a link of the catering chain system from Taiwan, is a restaurant of the Western cuisine for exhibiting the classic elegance and upper-class flavor. The signboard in carmine and golden colors is in the standard color of the brand and attracts the vision of the eyes of the passers-by at the most outstanding location; and such attraction has always been the most important arrangement in design.

In design, the partitions have been used to demonstrate the difference between the surface image of dazzling prosperity and the aesthetic and neat luxury. Therefore, only with the spatial streamlines and division resulting from the cold treatment can exhibit the flavor of elegance without show of vulgarity in a luxurious environment. With the application of streamlined and bright artistic wall painted with sea waves, the arc-shaped marble staircase for connection with finite tension leading upwards and the metal-thread-included glass screens in dual colors of gold and silver, a kind of rhythmic sense ix continuously used to exhibit the elegance of the compartment space. The customers' seats have a great amount of diversified variations such as leather type seats, velvet-wrapped chairs with high backs, noble first-class cabin chairs and arc-shaped seats wrapped in purple-colored cloth. Under the illumination of lamp lights, the golden chains shine with crystal transparency. Among the customer seats, partitions of crafty cloth lanterns and soft curtains serve as the means for the purpose of privacy and quietness. In terms of the materials, the sound-absorbing elements have been greatly intensified for obtaining a very quiet, fine-quality catering environment.

Dazzling aluminum thread lamp strings and color-painted glass plates also became the materials for artistic exhibition. The washroom culture is a link of great emphasis in the catering space. The arrangement with the use of golden glass wall, compound type table surface and gold-framed door cases with great nobility give out the neo-classic romance with the interpolation of classic red and golden colors.

北京王品西单店

　　王品展店中坪效最高格局最方正的店。因位于商务办公楼的下沉式广场中，将店招争取到街廓面，使得40m×70cm的LOGO煞是亮眼。弧形的空间分割是由柜台处左右回旋划分而成的自由曲线。

The arrangement with the highest plateau efficiency in the Wang Pin Restaurant rests with the typically square points. Owing to its location in a sunken type square of the commercial office building, strenuous efforts were made to set the shop signboard on the street front surface, resulting in the extremely attractive LOGO with the size of 40m × 70cm. The arc-shaped division of the space led to the formation of a free curve rotating from the left and right sides of the service counter.

主 设 计：张秀贞
参与人员：关中杰、赵立
摄 影 师：周宇贤
主要材料：大理石、云彩石、茶镜、沙金玻璃、布艺灯笼
坐落地点：西城区置地星座
面　　积：852m²
完工时间：2005.05.08
协力厂商：上海高格建筑装潢

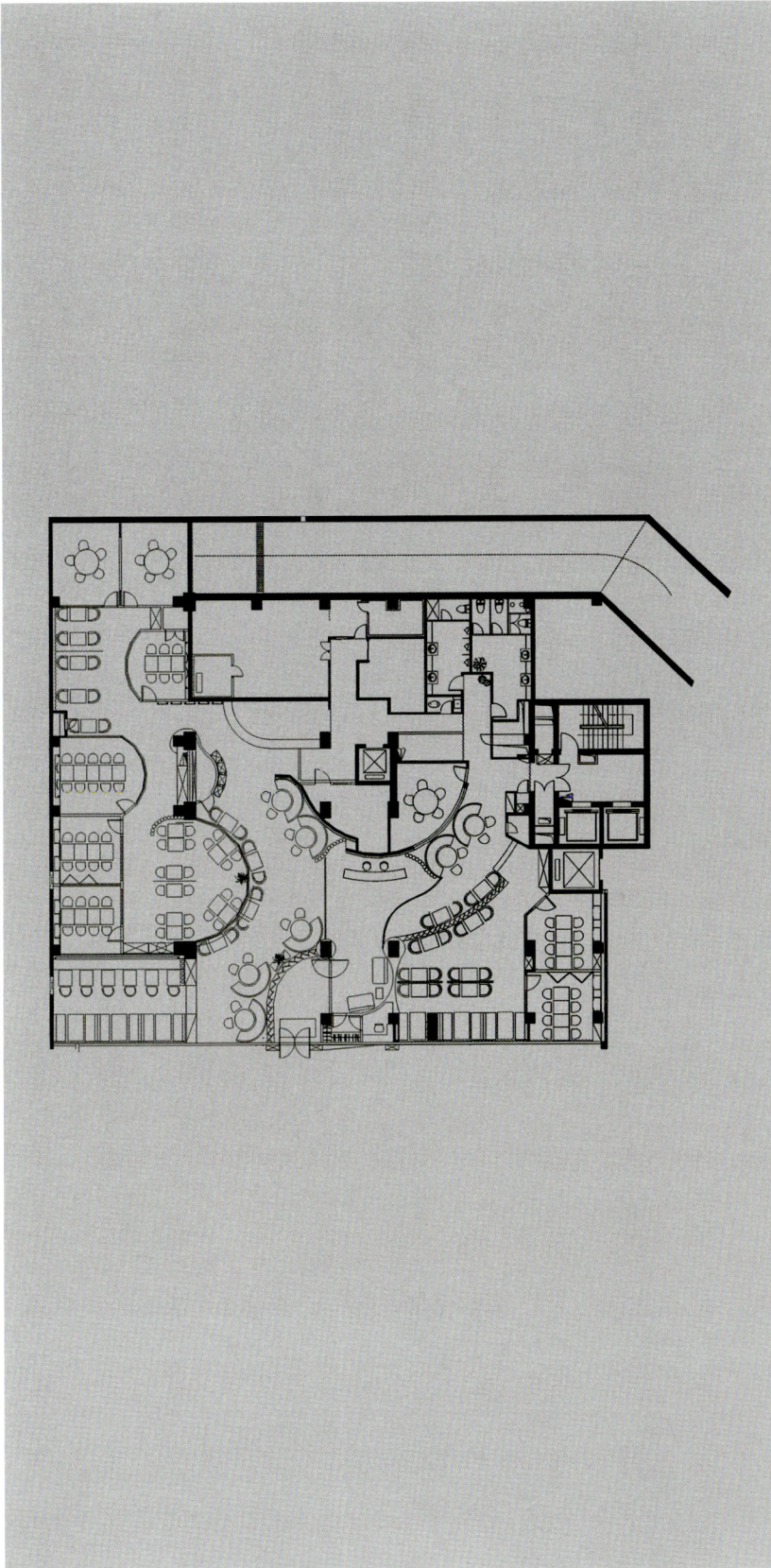

北京王品国贸店

在一层的平面中安置近百位客席，弧形卡座与高背情人椅等，在没有窗户的地下室中，巧妙的互动为彼此的景致，其间还包括两个活化的水景设计。

Nearly 100 customer seats have been arranged on one single floor. In a basement without windows, the arc-shaped seats, lovers' chairs with high backs and so on interact as one another's attractive views meticulously with two activated water-view designs being included among them.

主　设　计：张秀贞
参与人员：关中杰、赵立
摄　影　师：关中杰、韦培
主要材料：大理石、云彩石、茶镜、沙金玻璃、布艺灯笼
坐落地点：北京市朝阳区建外永安里
面　　　积：1200m²
完工时间：2004.02.15
协力厂商：北京柏易装饰

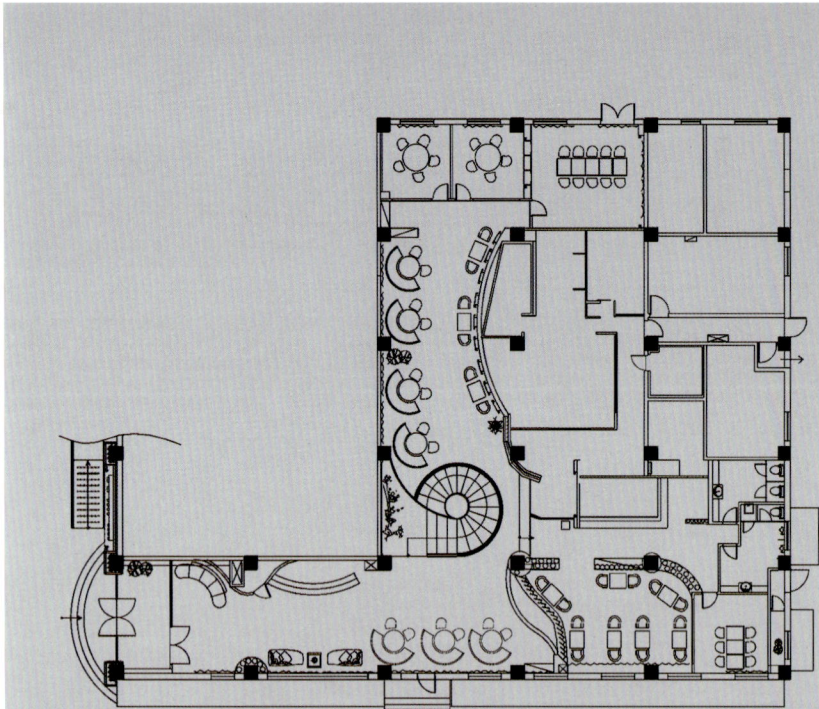

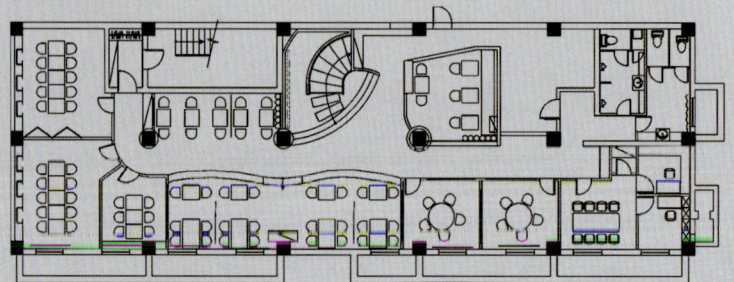

苏州王品河滨店

　　位于独栋具质感的建筑一、二楼，外观、云石、招牌同样是设计的焦点。采光入口、豪华楼梯是空间点睛之处；双层厨房开创工程管理上的突破记录；充分利用窗景，包房的设计便有了处处是景的商务用餐空间。

It is located on the first and second floor of a detached building with the textural sense. The appearance, marble stones and signboard are similarly the focal point of the design. The lighting entrance and the luxurious staircase serve as the eye-catching spots for the space. The double-layered kitchen broke the record in the engineering management. With the full use made of the landscape views out of the windows, the design of the compartments results in a commercial catering space with a belle view at any location.

设 计 师：张秀贞
参与人员：关中杰、赵立
摄 影 师：莫尚勤
主要材料：大理石、云彩石、茶镜、沙金玻璃、布艺灯笼
坐落地点：苏州市工业区湖滨新大地
面　　积：1183m²
完工时间：2005.07.01
协力厂商：上海海银建筑装饰工程

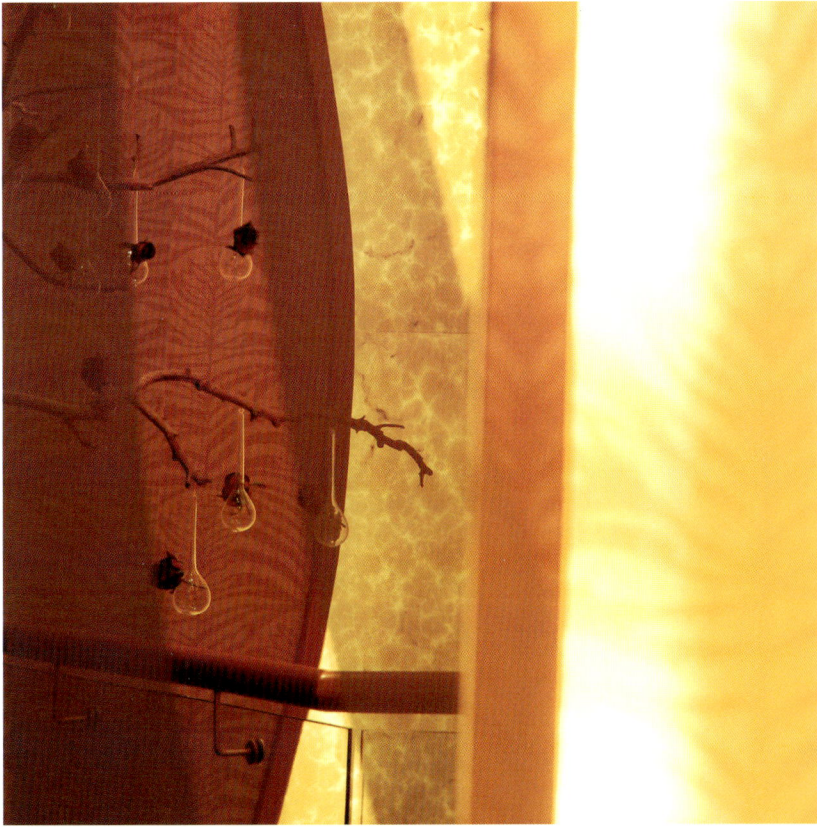

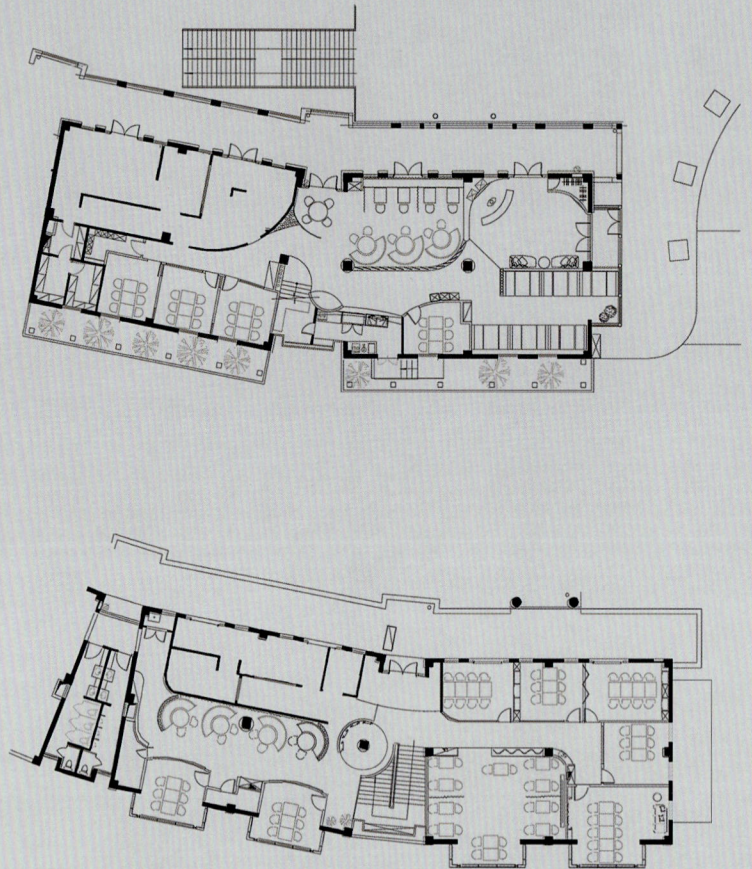

上海王品仙霞店

豪华的户外场景、创新的一楼空间、与旋转而上弧型梯，搭配整体新古典优雅的设计风格，是本连锁体系门店对其奢华贵族风明确定位的伊始。

The luxurious outdoor scenes, innovative ground-floor space with seats and the arc-shaped rotary staircase plus the design style of ensemble neo-classic elegance in coordination constitute a start point for the clearly defined positioning of the aristocratic style for the present chain shop.

主　设　计：张秀贞
参与人员：关中杰、赵立
摄　影　师：刘全辉
主要材料：大理石、云彩石、茶镜、沙金玻璃、布艺灯笼
坐落地点：长宁区仙霞路
面　　　积：823m²
完工时间：2004.03.31
协力厂商：上海高格建筑装潢

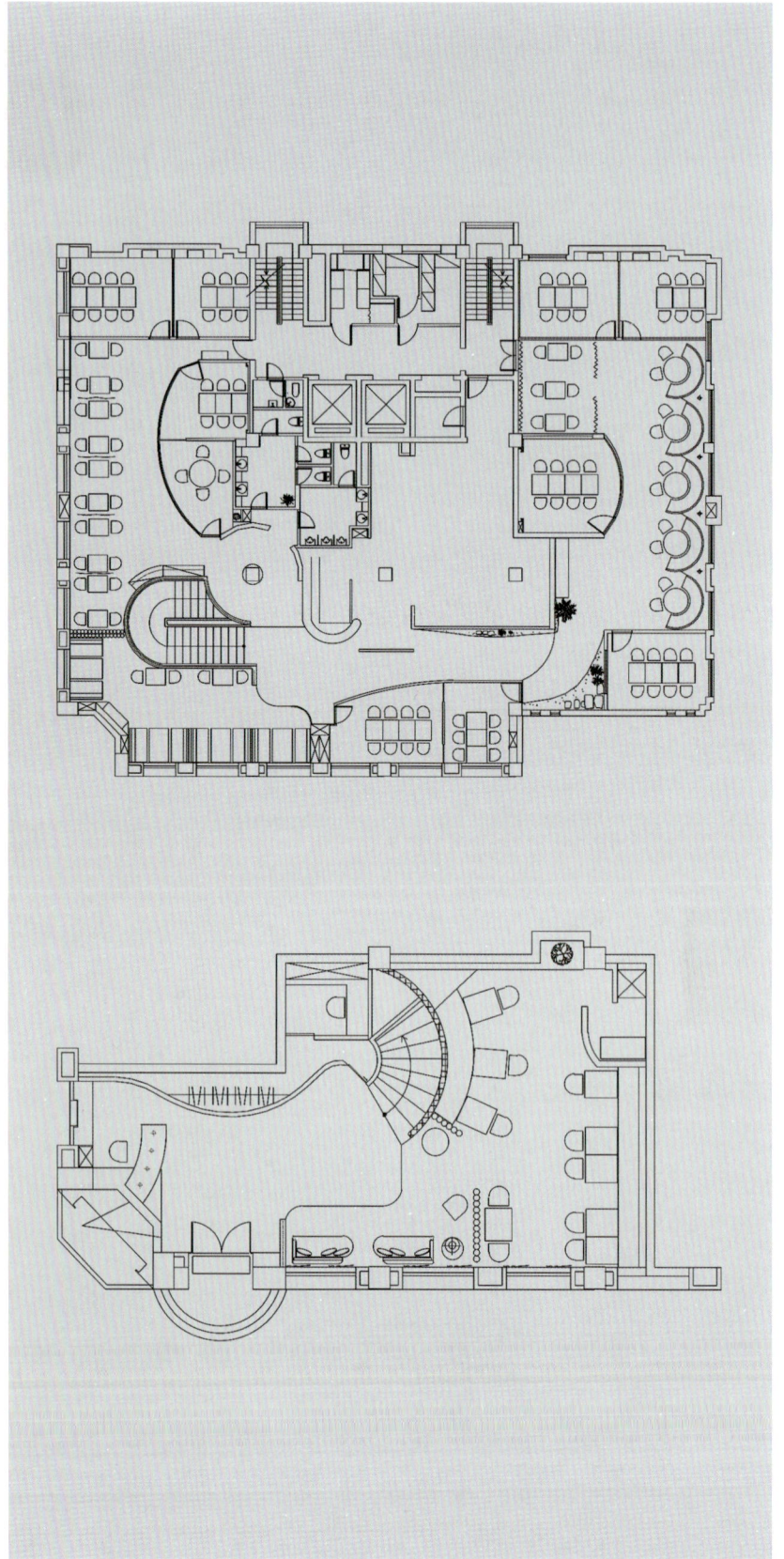

上海王品长乐店

　　本案位于迪生商厦的二层，因商厦经营升级，将整合原七个商场区块为一。原有的多根柱子与管线，因为软包的设计运用而不见了。红酒柜是本案的设计焦点，在低照度的光线中表现餐厅的高格调。并在封闭的空间中大量假造落地窗景，环绕的动线规划，也是不同功能空间的引导轴。

　　The present project is located on a grand second floor of Disen Mansion. As a result of upgrading the business operation in the commercial mansion, the original seven commercial blocks were integrated into one block. The originally existing multiple pillars and pipelines disappeared with the application of the soft-wrapping design. The red wine cabinet served as the focal point of the design of the present project, which exhibited a superior artistic style of the restaurant. The artificial French window views were made in an enclosed space. The encircling moving-line planning also served as a guiding axle for the different functional areas.

主 设 计：张秀贞

参与人员：羊山木　赵立　ERIC

摄 影 师：草尚勒

主要材料：大理石、云彩石、茶镜、沙金玻璃、布艺灯笼

坐落地点：卢湾区长乐路

面　　积：719m²

完工时间：2005.12.23

协力厂商：上海高格建筑装潢

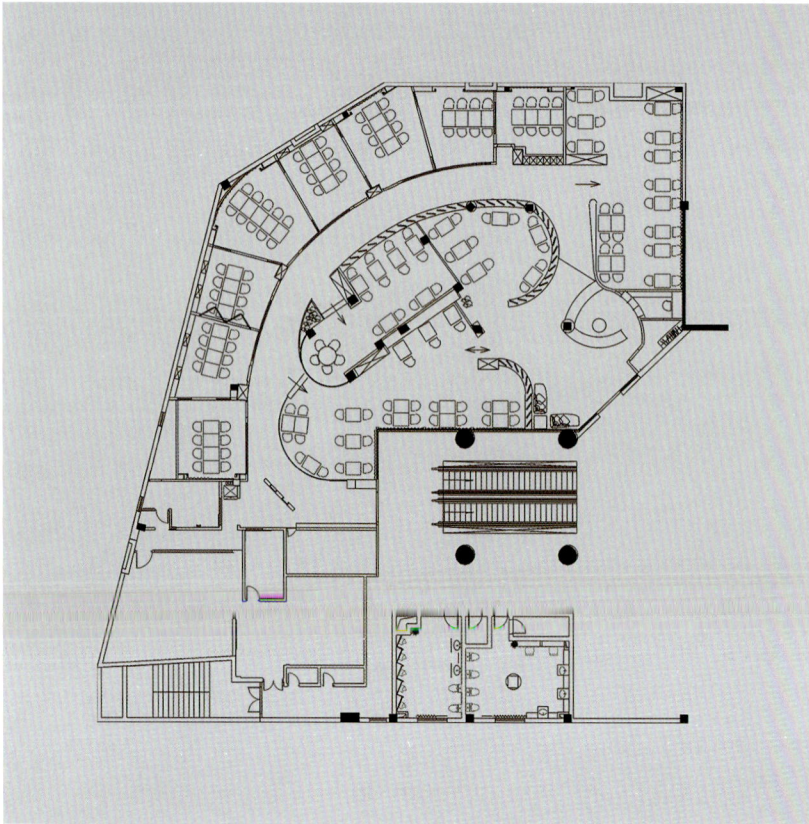

深圳王品深南店

　　重新创造的楼梯，给予商业空间一个活化的动力。以弧线格局化解柱子多的空间尴尬；虽然营业面积为本系列中最大，但接受优越城市文化洗礼，客席与营运置放最宽尺度，水晶装饰细节仍可体现新古典玩味。

The recreated staircase gives an activated motive power to the commercial space. The arc-shaped arrangement has dissolved the spatial dilemma with too many columns. Though the business operating area is of the largest in the present series, yet with the acceptance of the baptism of a superior urban culture, the arrangement of the customer seats and operating devices has been given the largest dimensions. The details of the crystal decoration may still exhibit the neo-classic recreational flavor.

主 设 计：张秀贞

参与人员：关中杰、赵立

摄 影 师：羊中木、丰楷

主要材料：大理石、云彩石、茶镜、沙金玻璃、布艺灯笼

坐落地点：福田区深南中路

面　　积：1663m^2

完工时间：2005.08.31

协力厂商：上海高格建筑装潢

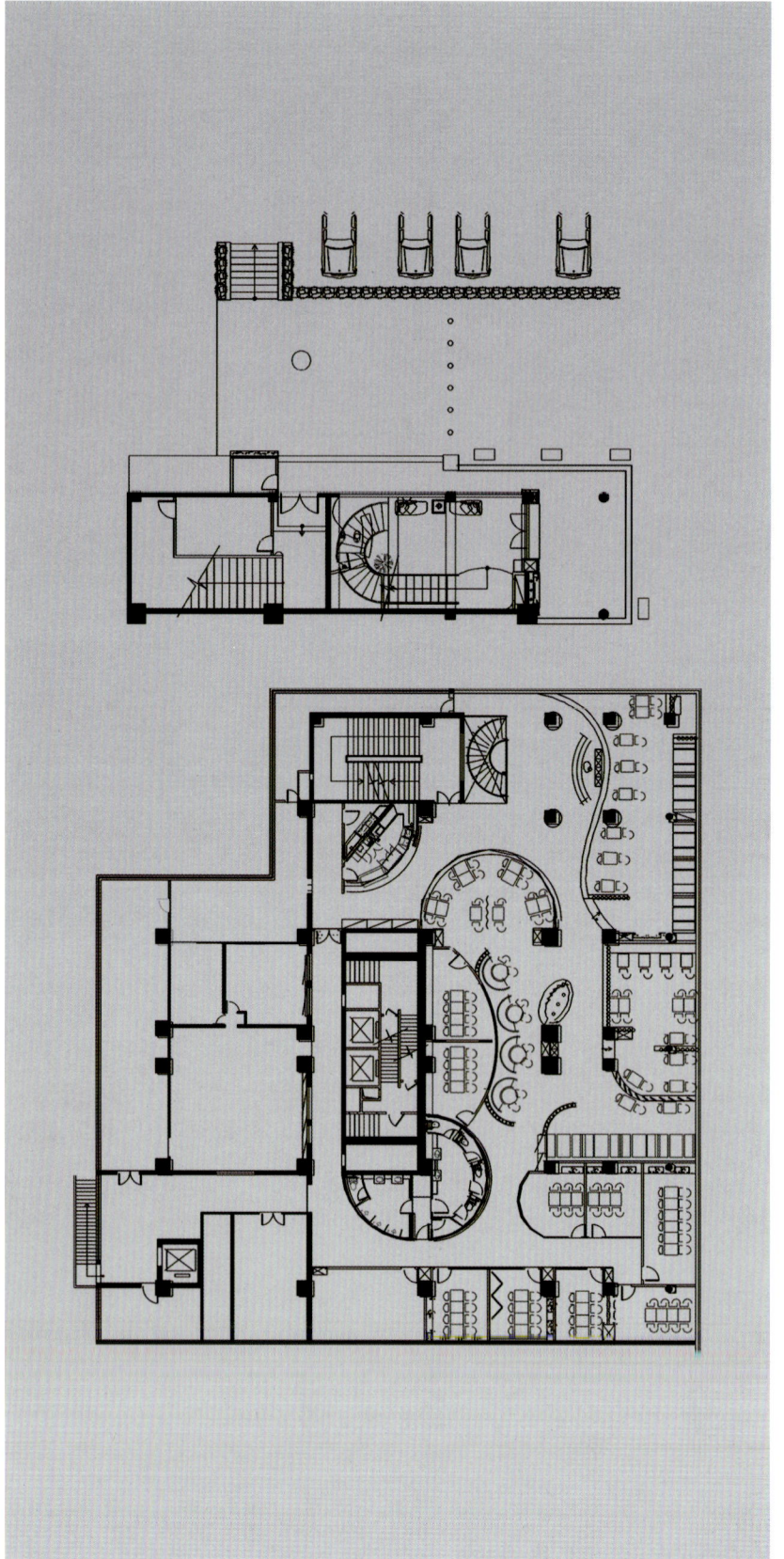

唐吉诃德的骑士精神

上海豆捞坊

时尚风格下的狂野热情需要洋红、深灰与不锈钢的亮丽来表现。大量的中国红剪纸艺术变化及立体铁花艺术墙，均是在红色上做不同深浅曲度圆形的表现。水泥漆板经过再度上漆后的安定性格，有着青石般的质感。

The frantic enthusiasm under the fashionable style requires the colors of carmine and dark grey and the brightness of stainless steel for its expression. A large amount of variations of the Chinese red paper cuts are based on the expression of the red color of the different tones and the circles with the different curvatures. The stabilized character of the painted cement boards following the re-painting has the textural sense of bluestones.

上海豆捞坊浦东店

　　快意感的释放被用来诠释本案。浅蓝色的恣意、红与黑的浓色布置，让主廊道的展示尽显现代与趣味感。视线末端的酒杯灯装置、圆心柱、红纱幔、红灯笼，都将热情的快意感释放出来。灯光则利用阴暗处做空间的区隔，虚拟的神秘不经意也漫延出来。

　　The releasing with the sense of joyfulness has been used for explaining the present project. The willfulness of the blue color and the strong-colored arrangement with the red and black colors enable the demonstration in the main corridor to show the sense of modernity and tastefulness. The wine-cup-shaped lamp device at the end of the visual sight, the concentric pillars, the red gauze curtains and the red lanterns all release the enthusiastic sense of joyfulness. As for the lamp lights, the dark locations have been used for making spatial partitions, thus diffusing a kind of fictitious mystery unintentionally.

主 设 计：张秀贞
参与人员：关中杰、叔立
摄 影 师：莫尚勤
主要材料：玻化砖、不锈钢、水泥板、胡桃木线条、夹丝布玻璃
坐落地点：浦东新区张杨路
面　　积：1400m²
完工时间：2006.05.15
协力厂商：上海柏思艺术装潢设计

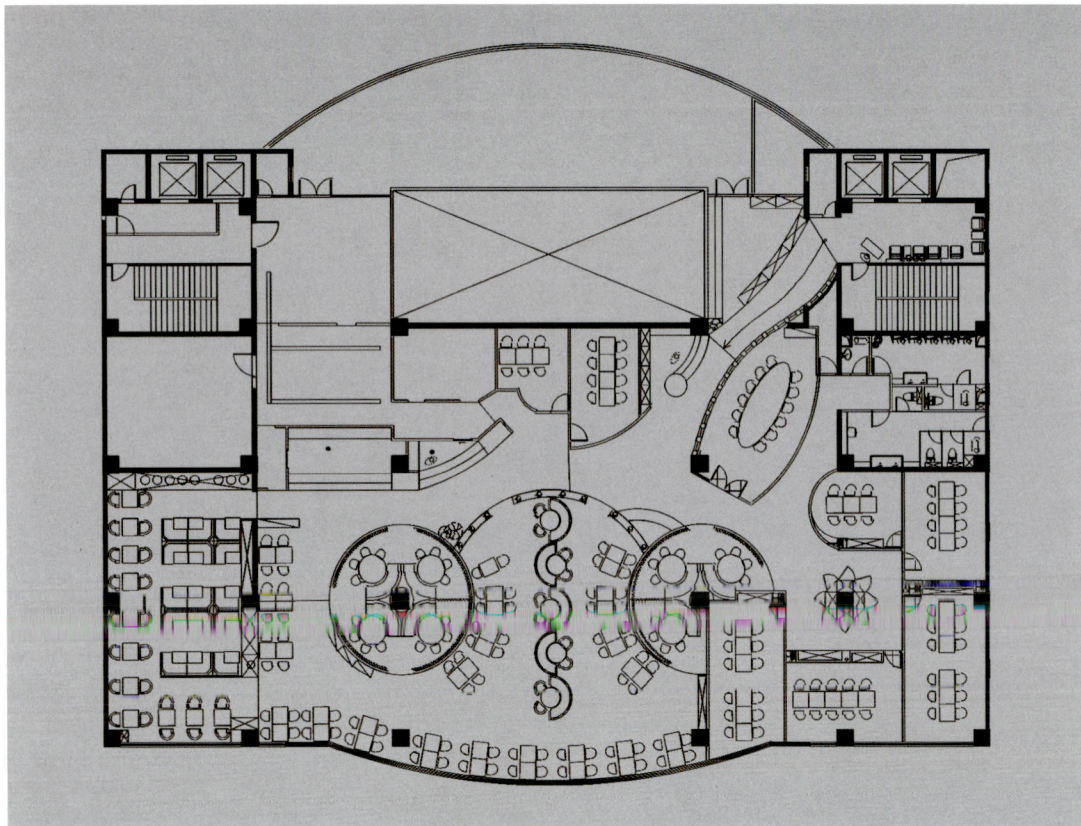

上海豆捞坊百乐门店

入口的两块意大利进口的彩流板是一个趣味迎宾的设计。接着看见一排方形玻璃红柱，创造了一种视觉连续感的自由曲度空间感。红酒吧台、红色折合玻璃，再度营造视觉上的刺激。因位处静安寺美丽的夜景区，因此灯光设计上采用减少漫光的豆胆灯直接投射至桌面，顶楼的阳光屋靠窗座位以烛光照明，都是为保留户外夜景的考虑。

Two colored screen boards imported from Italy at the entrance is an interesting design for greeting the guests. What follows is a row of square glass columns in red color, thus creating a kind of spatial sense with free curvature of continuous vision. The red bar counter and the red folded glass pieces again build up the stimulation for the visual sense. As the restaurant is located in a night scenic area at Qing'an Temple, the design of lamp lights employs a kind of lamp with less diffused light for direct reflection onto the table surfaces. The seats alongside the windows on the attic facing the sunshine are lighted with the candle lights out of the consideration of reserving the outdoor night scenes.

主 设 计：张秀贞

参与人员：关中杰、赵立

摄 影 师：吴尚勤

主要材料：玻化砖、不锈钢、水泥板、胡桃木线条、夹丝布玻璃

坐落地点：卢湾区南京西路

面　　积：949m²

完工时间：2006.04.01

协力厂商：上海高格建筑装潢

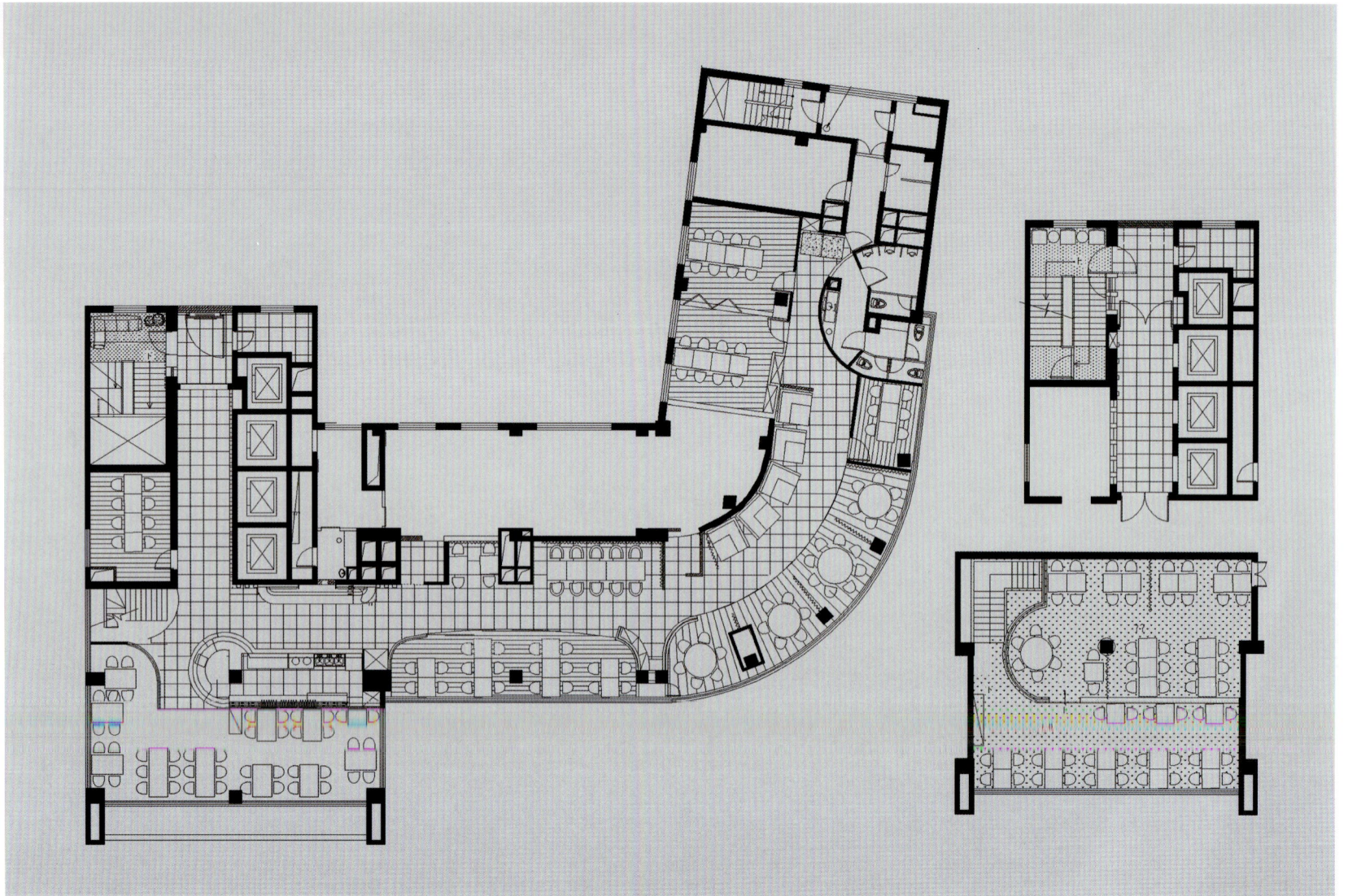

豆捞坊中山店

　　本案是豆捞坊的豪华版设计，从面积、座位层次、虎豹纹地毯，即刻意营造的Lounge的感觉，无一不显现大气而豪华的排场。火锅常用的肉片卷被形象为装饰的符号，运用铁艺做成卷曲状双夹于玻璃内，是一种对材料的挑战与成功尝试。

The present project is a design of the luxurious edition for the Doulaofang Restaurant. From the size of the area and the seat layers to the carpets with leopard spots, painstaking efforts have been made to build up the feeling of a lounge; and they all display the ostentation and extravagance of magnificence and luxury. The image of the meat rolls which are frequently used for making chafing dishes has been used as the decorative symbols and built in the glass pieces as the included curves with the use of iron handicraft, which is a challenge to the material and also a successful trial.

主 设 计：张秀贞
参与人员：关中杰、赵立
摄 影 师：周宇贤
主要材料：LED灯，水泥板刷清漆，明镜，壁纸，软包，虎豹纹壁布，红黑丝线
坐落地点：上海长宁区长宁路
面　　积：800m²
完工时间：2006.11.30
协力厂商：大连钱柜阿兴建筑装饰工程公司

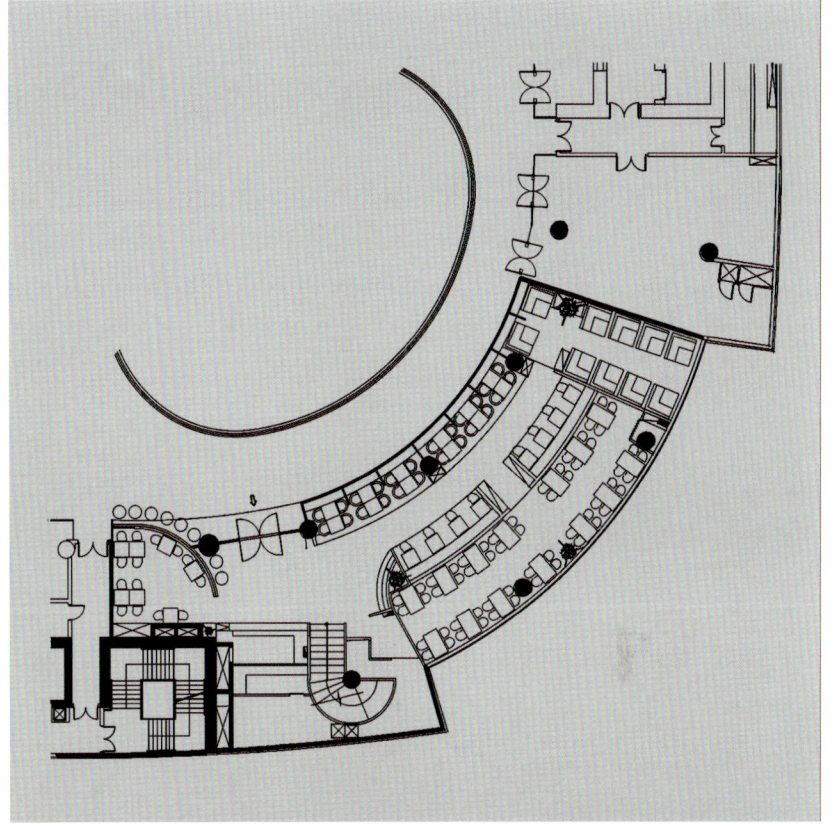
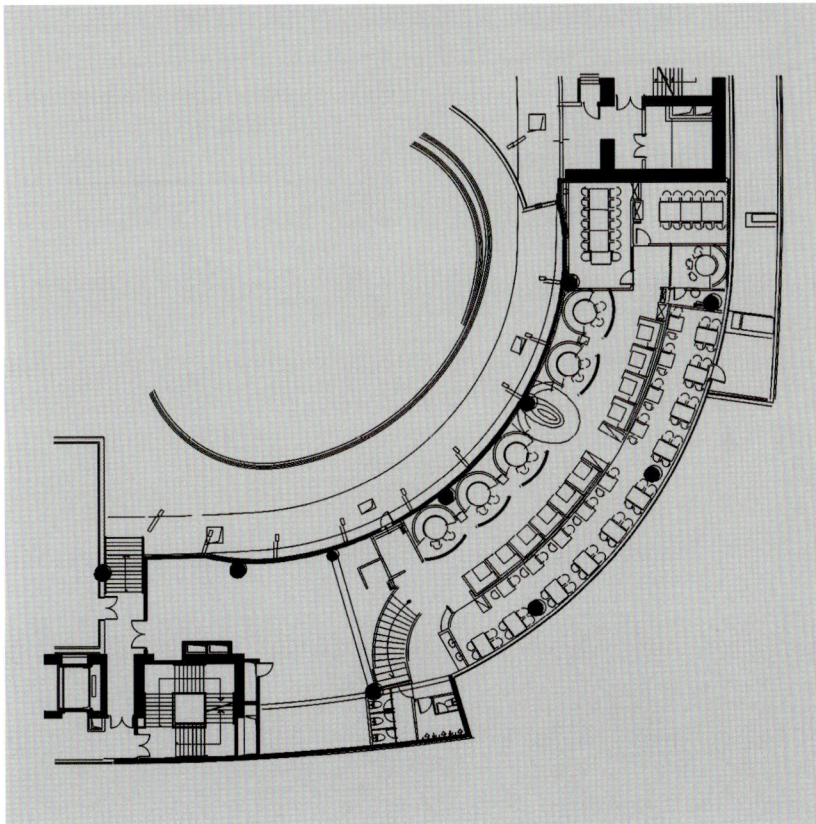

人文主义的中国风

丰滑火锅

在视觉印象中火鹤图腾是本案的表现重点，并且以人文主义风格为餐厅基调。将中国风以现代感呈现在装饰中，毛笔恣意挥毫的洒脱成就了一盏盏的方形灯笼。色调上红与金的洋洋洒洒倾泻，浓郁的中国风与欢乐气氛被塑造至最高潮。另外在火锅餐厅的高温空间中，如何将新鲜空气送入与将热气排出，使得食物气味不交杂，是考验工程管理的最大挑战所在。

In the visual impression, the fire crane totem is the focal point of expression in the present project; and the humanistic wind serves as the basic tone of the restaurant. With Chinese wind exhibited in the decoration with the sense of modernity, the free and easy style shown by wielding of a Chinese brush led to the depiction of a series of square-shaped lanterns. In terms of colors, the free damping of the red and golden colors plastered the strong Chinese wind and the merry atmosphere to the highest tide.

苏州丰滑狮山店

　　自由取食的食物台，犹如西餐的水吧台，鲜花与水景达成了视觉及听觉的飨宴。扇形与半开放设计的座位区，夹布玻璃与金色珠帘，因量化而不断延续着视觉。另外在火锅餐厅的高温空间中，如何将新鲜空气送入与将热气排出，使得食物气味不交杂，是考验工程管理的最大挑战所在。

　　The food table for free taking of foodstuff is just like the water counter in Western cuisine. Fresh flowers and water scene constitute a delicious feast of both visual and hearing sense. The seat area with the fan-shaped and semi-open design, the glass pieces with inclusion of cloth and golden pearl curtains continuously prolong the vision because of quantization. Besides, in the high-temperature space of a chafing dish restaurant, how to deliver fresh air and expel the hot air for no mixing of smells for food offers the greatest challenge to the engineering management.

主　设　计：张秀贞
参与人员：黄中杰、赵立
摄　影　帅：员向瞰
主要材料：清玻璃夹布、防火板、玻化砖、玻璃马赛克、金属链
坐落地点：苏州市狮山路 35 号 金河大厦
面　　　积：799m²
完工时间：2006.05.25
协力厂商：上海高格建筑装潢

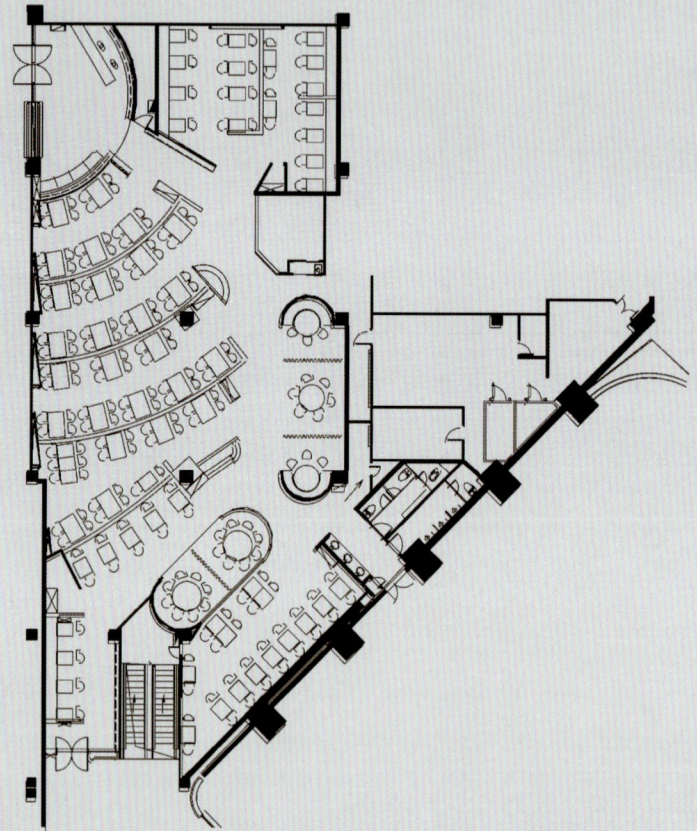

上海丰滑中环百联店

在 L 形的平面中，本案最成功之处是对空间区隔的规划。镜子在转角处发挥了空间延放的效果。入口利用转角面做了大气的场景设计，一盏红罩灯与水池呼应，夹布玻璃隔断巧妙的区分了内外空间。

In terms of the L-shaped plane, the greatest success for the present project rests with the planning of the spatial partition. Mirrors have given play to the effect of spatial extension. At the entrance, the corner-turning surfaces have been used for making a grand scenic design. A chain of lamps with red shades corresponded with a water pond; and the cloth-included glass partitions meticulously separated the inner and outer spaces.

主 设 计：张秀贞
参与人员：关中杰、赵立
摄 影 师：周宇贤
主要材料：玻化砖，啡网纹大理石，壁纸，明镜，墨镜，夹布玻璃，裂纹漆，珠链，羊皮纸灯笼
坐落地点：上海普陀区真光路百联中环购物中心
面　　积：680m²
完工时间：2006.12.18
协力厂商：上海华亮建设发展有限公司

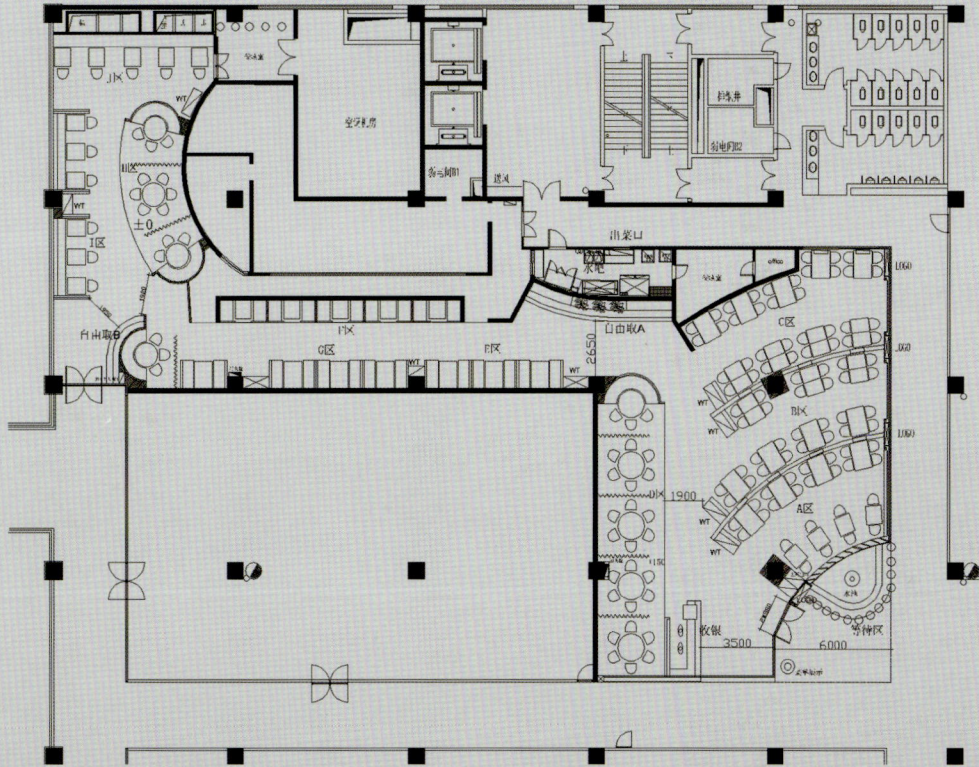

莫里斯的神秘电影

悦唐时代火锅

　　同样的在临街面塑造一焦点门面——20m×10m 的转角设计。由一楼至地下层楼梯悬浮式斜坡灯光装置，是一华丽动线的延伸，及室内无数的吊灯辉煌，皆宛如歌剧院的盛大与庄重。时尚的元素俯拾皆是：如黑色金属地砖与冷色系钢索线共同构成接待台的主墙；背投光银丝柱面、气泡水柱的律动感与水声琅琅。

Similarly, a shop front was built on the side facing a street with the corner-turning design of 20m × 10m. The lamp light arrangement of the suspension type ramp along the staircase from the first floor to the basement is an extension of a beautiful moving line. The brilliance formed by numerous hanging lights indoors appears just like the magnificence and grandeur of an opera theatre. The fashionable elements spread here and there such as the black metallic floor tiles, the main wall composed with the steel ropes in cold colors adjoining the reception counter, silver thread column surfaces with backward light reflection and the sense of rhyme and the flowing sound of water resulting from the water bubble columns.

主 设 计：张秀贞

参与人员：关中杰、赵立

摄 影 师：关中杰、韦培

主要材料：大理石、夹层纱、清玻璃夹金属丝、不锈钢马赛克

坐落地点：朝阳区大望路光华路口

面 积：920m²

完工时间：2005.12.23

协力厂商：上海美达建筑工程有限公司

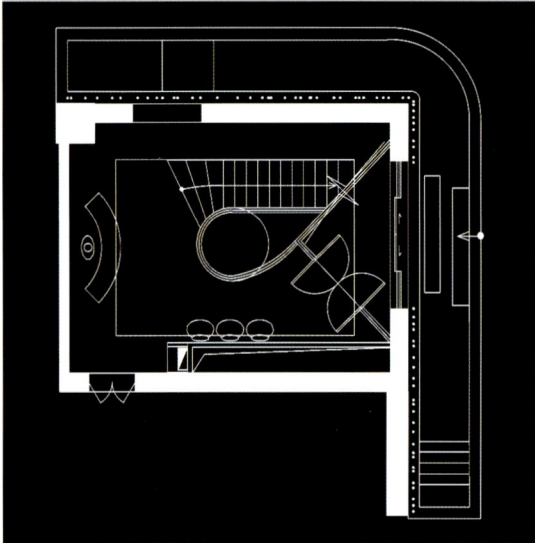

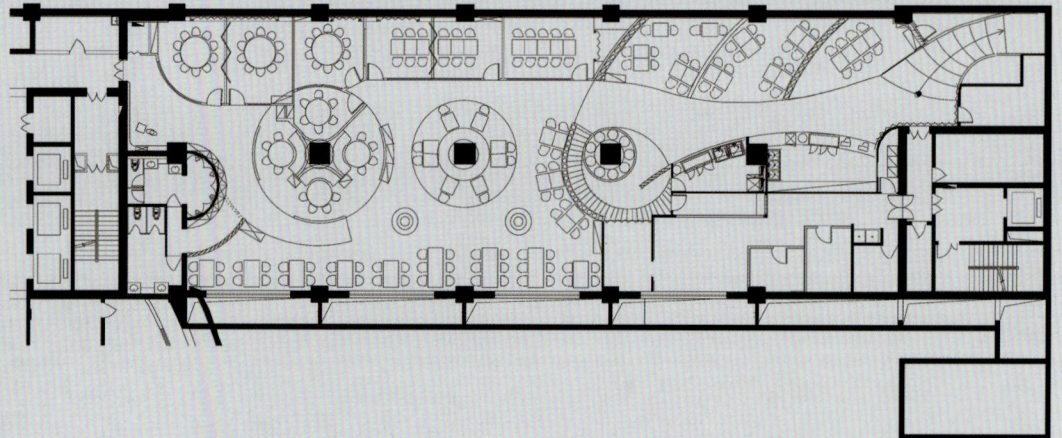

放肆的情人

789 新概念火锅餐厅

"789 餐厅"的命名有继往开来的风水运势含意，是新世纪时代所独具的现代感。餐厅主张好锅好底并诉求于真材实料，墙上似博物馆般以玻璃框盒展示着火锅锅底的真药材。也因为此真实本性的坚持，延展设计材质运用了大量的自然元素为装饰符号，石头、干花草、玻璃、不锈钢等犹如风情万种的情人，以冷灰色调混合着热情的红色，恣意的释放着情感，但那份放肆却又流露着真性情，一切显得质朴而自然。可秀吧设计亮点为精致厨艺，放肆出钻石级钟情，不锈钢球垂空而下，如镜子般反射出全区大千世界。本案最成功处在于控制材料成本的情况下，依然能将设计情感完整释放表现出来。

The name of "789 Restaurant" has the connotation of the operating trend of geomancy for carrying on the past tradition and opening a way for the future and bears the unique sense of modernity featured by the epoch of the new century. The restaurant advocates the principle of good pot and good base materials for chafing dishes and claims for the use of genuine materials. The glass frames or boxes fit on the wall exhibit the genuine herbal materials used as the base for the pot. Owing to the insistence on the inborn nature of genuineness, the designed material textures have applied a large amount of natural elements as the decorative symbols such as stones, dried flowers and plants, glass, stainless steel and so on just like a charming lover. The cold grey color has been mixed with the enthusiastic red color for releasing the emotion recklessly. Yet the recklessness betrays a kind of genuine temperament. All reveals the simplicity and naturalness. The bright spot of the bar design rests with the exquisite cooking art for the reckless release of the diamond grade love. The stainless steel balls are hanging downward from the ceiling for reflecting the kaleidoscopic world of the entire area. The most successful point of the present project rests with the full release and expression of the entire design emotions under the conditions of controlling the cost of the materials.

主 设 计：张秀贞

参与人员：关中杰、赵立

摄 影 帅：施昆城

主要材料：红银丝线，红色干枝，弥尔板，磨姑石，有色玻璃，河床石，不锈钢球

坐落地点：苏州工业园区湖滨大道

面   积：780m²

完工时间：2006.12.8

协力厂商：上海海银建筑装修有限公司

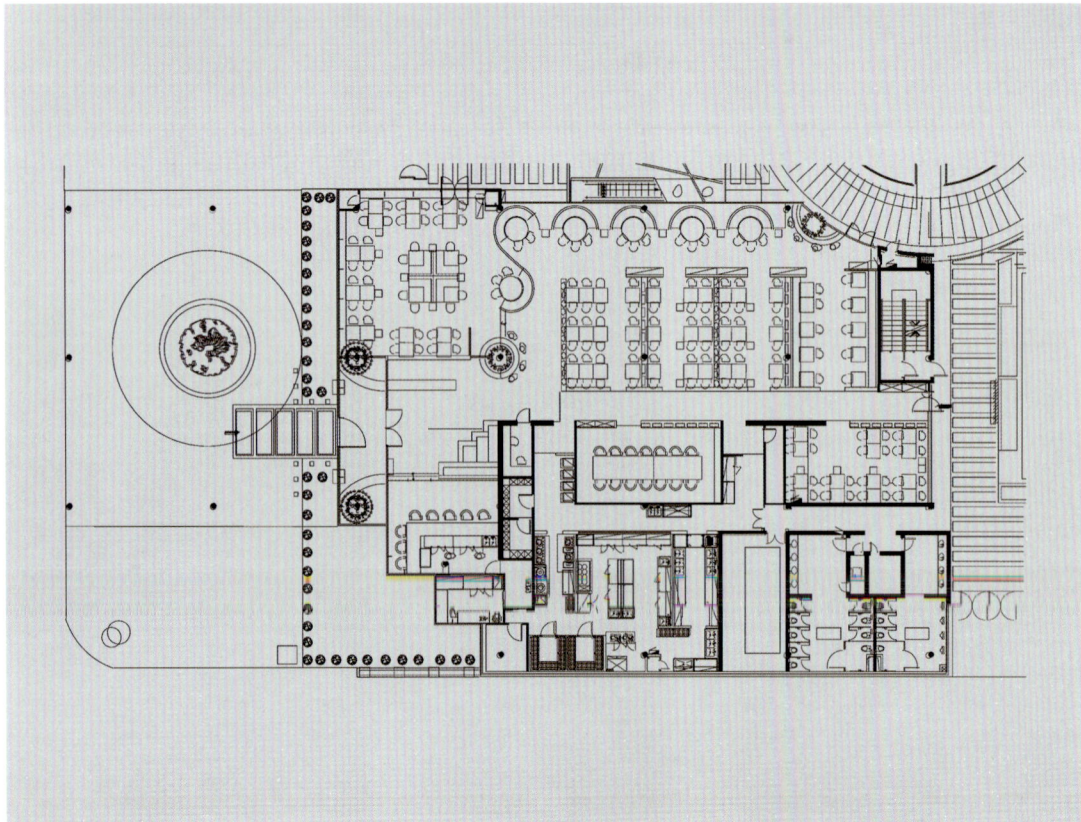

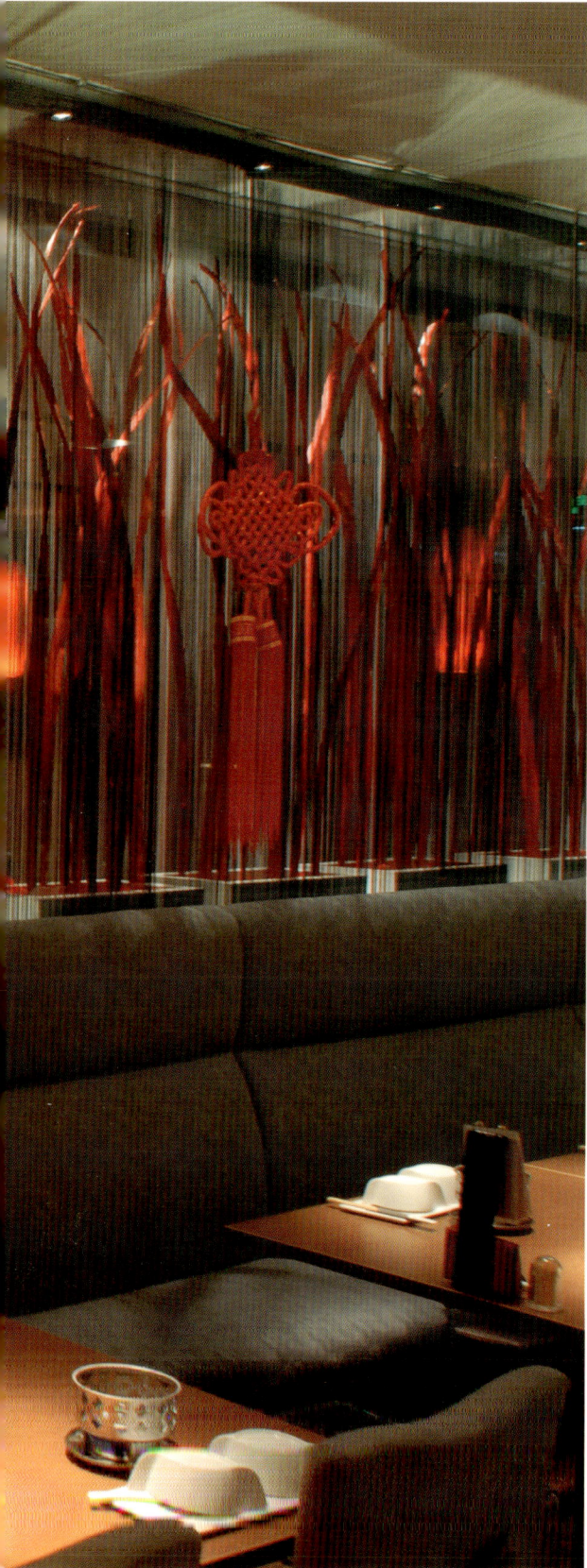

樱花树下的禅风

陶板屋日式怀石料理

怀石料理为日式庭园与生活艺术综合的餐饮文化，以陶板艺术为设计主流，辅以池泉庭园的和风主题餐厅。

池泉，欲将大自然山水集揽于一室，遐想在樱花树下惬意与友邀月欢饮。陶板屋的5.5米挑空是空野聚会最佳环境。巧妙的客席区隔，突破入口与大厅1米高度落差的问题。意象钢索、漫波水景、灯笼的古朴美，柔化了结构的线条，舒适的弧梯、虚实的造景石墙及抽象三裸女图布置，都是整场的重点。VIP尊客均在各层次中，依序流动，处处有景可观。连洗手间也有艺妓画作，处处惊艳。其精致、自然均衡的构图和化繁为简的设计手法，甚得欣赏者喜爱。另为配合各种饮食形态，活动竹帘成了最好屏障。170个座席，仍维持处处为VIP的尊贵，上、下夹层各具风格，有隐私亦有开放。我们把每个窗景更深入装饰，犹如在贵族禅风庭院中享用一道道艺术佳肴。

Huaishi Restaurant offers a catering culture with the integration of the Japanese style courtyard and living arts and employs the ceramic board art as the mainstream of design with the supplement of a courtyard featured with water ponds and fountains and a canteen with the gentle breeze as the theme.

With the encountering of water ponds and fountains, one tries to concentrate the mountains and rivers of Nature into one house and intends to drink wine joyfully with the friends under the oriental cherry trees. The ceramic board house (Taobanwo) with the cantilevering of 5m serves as the best place for gathering the stars on an open wilderness. The clever partition of the customer seats found a solution to the problem of difference in height for 1m. The imaginary steel ropes, the wavy water scene and the simplistic beauty of the lanterns have softened the lines of the structure. The comfortable arc-shaped staircase, the landscape-like stone walls in fiction and the arrangement of a painting depicting three naked ladies serve as the focal point of the entire scene. The VIP customers may scatter on various floors and move in good order for viewing the scenes wherever they are. Even in the washroom there are paintings of geisha girls so as to leave beauty at every place. The delicate, natural and balanced configuration and the design technique of turning complexity into simplicity have received warm response from the appreciators. For coordinating with various types of catering morphology, the movable bamboo curtains become the best screens. A total of 170 seats maintain the nobility for all the VIP customers. The upper and lower intercalations have the different flavors and offer either privacy or openness. The in-depth decorations have been made for every window sight just as enjoying every delicious dish in an aristocratic courtyard with the Chan style.

主　设　计：张秀贞

参与人员：关中杰、赵立

摄　影　师：刘圣辉

主要材料：真石漆、钢索、仿马赛克地砖、铁刀木皮、芦苇卷帘、夹布玻璃、漆板玻璃

坐落地点：上海市肇家浜路枫林路口

面　　　积：242m²

完工时间：2003.12.31

协力厂商：上海海银建筑装饰工程公司

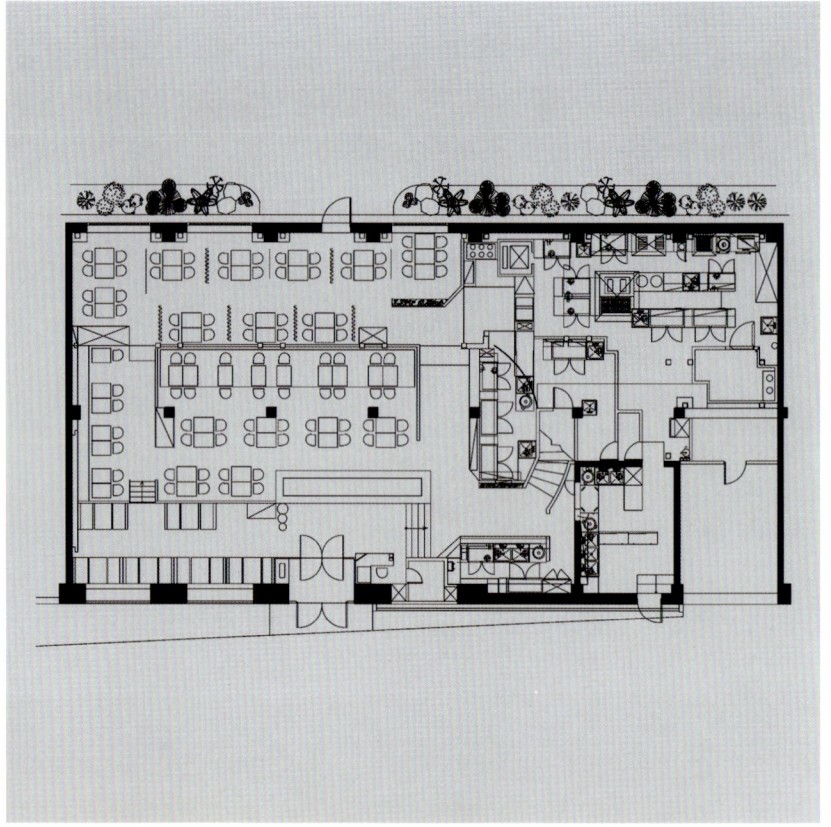

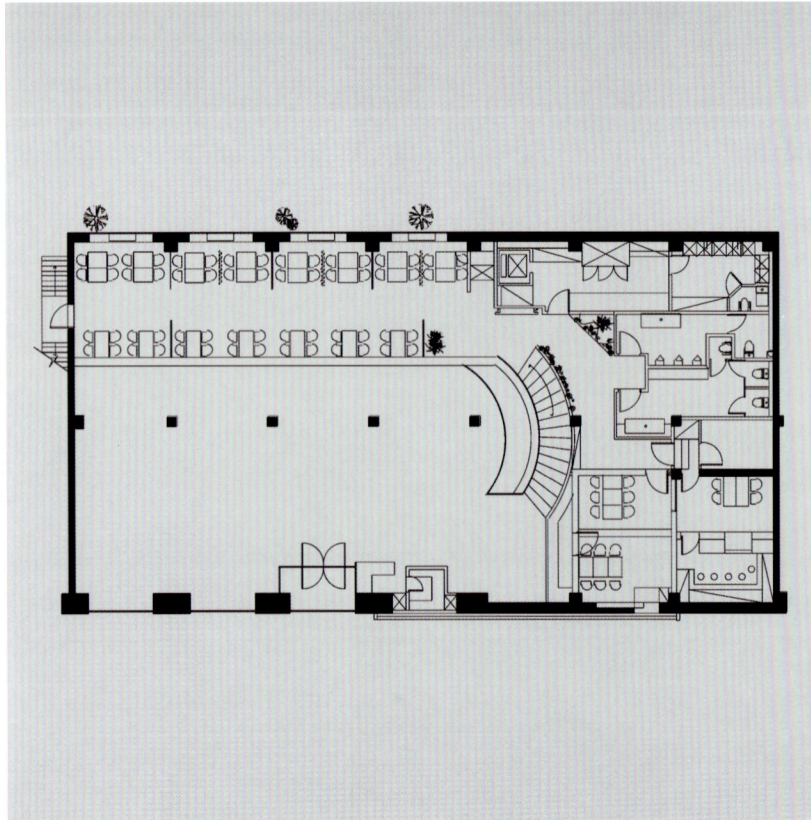

超现实的幻象世界

澜时尚料理

　　本案为幻象能量激活超现实般蓝色，装点出蓝镜、白线、透杯墙、桃光泛影等组件的转向，虚实之间精细地达到"不连接"的效果，于是时空便静静浮移于此，超自然与超现实，超真实情境流转，在于东西南北方向感的旋场。纵深开启这道门之后所有印象心境主宰、超时空投射出东方既出，超越生命力与美。顶层大都会景致于设计主轴。更添用餐丰富氛围。

　　设计者运用超现实涵构设计基调，与澜共体异位时空中 —— 时尚概念。

　　在室内空间中，体验大自然风、雨、阴、晴的宇宙多变幻象。

The present project involves the imaginary energy initiating the surrealistic blue color so as to bring out the directional turning of the blue mirrors, white lines, perforating cup walls, peach light shadow and some other components and achieve the effect of "non-connection" between the fantasy and realism; consequently, the temporal space quietly drifts here. The circulation of the supernatural, surrealistic and super-genuine situations rests with the rotary field of the directional sense of the east, west, south and north. Following the opening of this door longitudinally, the governor of all the impression type metal states will transcend the vitality and beauty with the super temporal space throwing out the ejected east. On the top layer, the metropolitan scenes become the main axle of the design, thus adding an abundant catering atmosphere.

The designer applied the surrealistic frame as the key tone of the design, which shares a common body with different position in the temporal space, i.e., the concept of fashion.

In the interior space, it is possible to experience the changeable illusions in the cosmos such as wind, rain, cloudiness and sunshine of Nature.

主　设　计：张秀贞

参与人员：关中杰、赵立

摄　影　师：周宇贤

主要材料：环氧树脂漆，LED 灯，银黑色丝线，酒杯立灯

坐落地点：上海武宁南路

面　　　积：900m²

完工时间：2006.10.8

协力厂商：上海高格建筑装潢

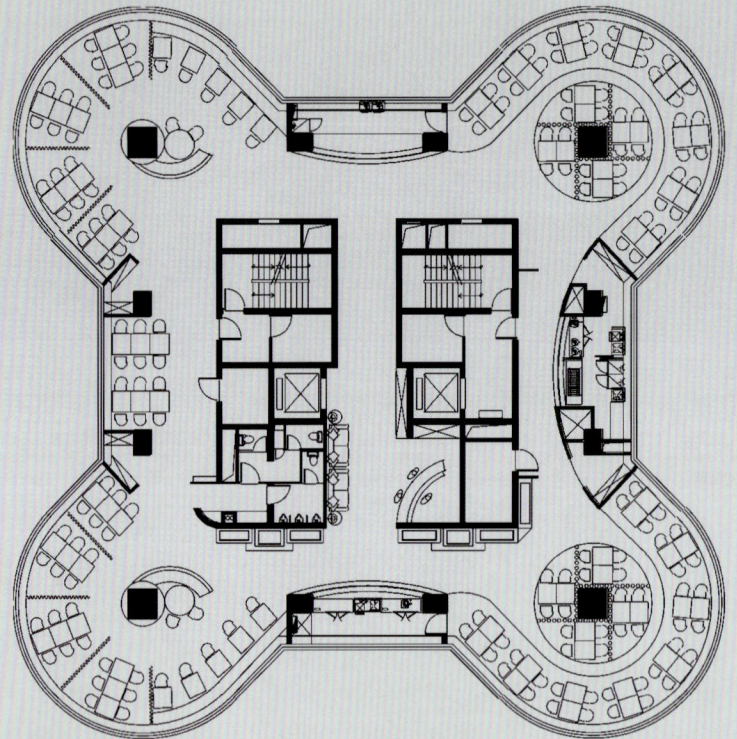

极简主义的法拉利

上海泰平天国

　　将设计放空至毫无装饰是本案设计的基调。高贵亮丽的东南亚热情则由橙色来铺陈，主墙的橙色有着泰式奶茶的浓郁与内敛。原色水泥板墙面、深灰色地板、墨镜墙壁共同塑造着冷静空间的气质。从狭长40m门面的入口折门至吧台，再至包厢，设计者完成了空间三个渐进层次的规划。橙色随着灯光的深浅漫射，也赋予了丰富的表情。

To vent the design up to non-existence of any decoration is the key tone of the design for the present project. The noble Southeast Asian type enthusiasm is unfolded through the use of the orange color. The orange color on the main wall contains the high concentration and inward convergence of the Thai type milky tea. The cement board wall surfaces in the original color, the dark grey floor boards and the black mirror wall surfaces jointly builds up a temperament of a cold space. From the turning at the entrance of the shop front of 40m in width to the bar counter and then to the chartered compartments, the designers completed the planning of three evolutionary layers of the space. The orange color diffuses either in dark or pale tones with the lamp light and also radiates the abundant expressions.

主　设　计：张秀贞

参与人员：关中杰、赵立、刘飞燕

摄　影　师：关中杰、韦培

主要材料：玻化砖、水泥丝板、人造石、三聚板

坐落地点：黄浦区南京西路

面　　　积：300m²

完工时间：2005.09.08

协力厂商：上海高格建筑装潢

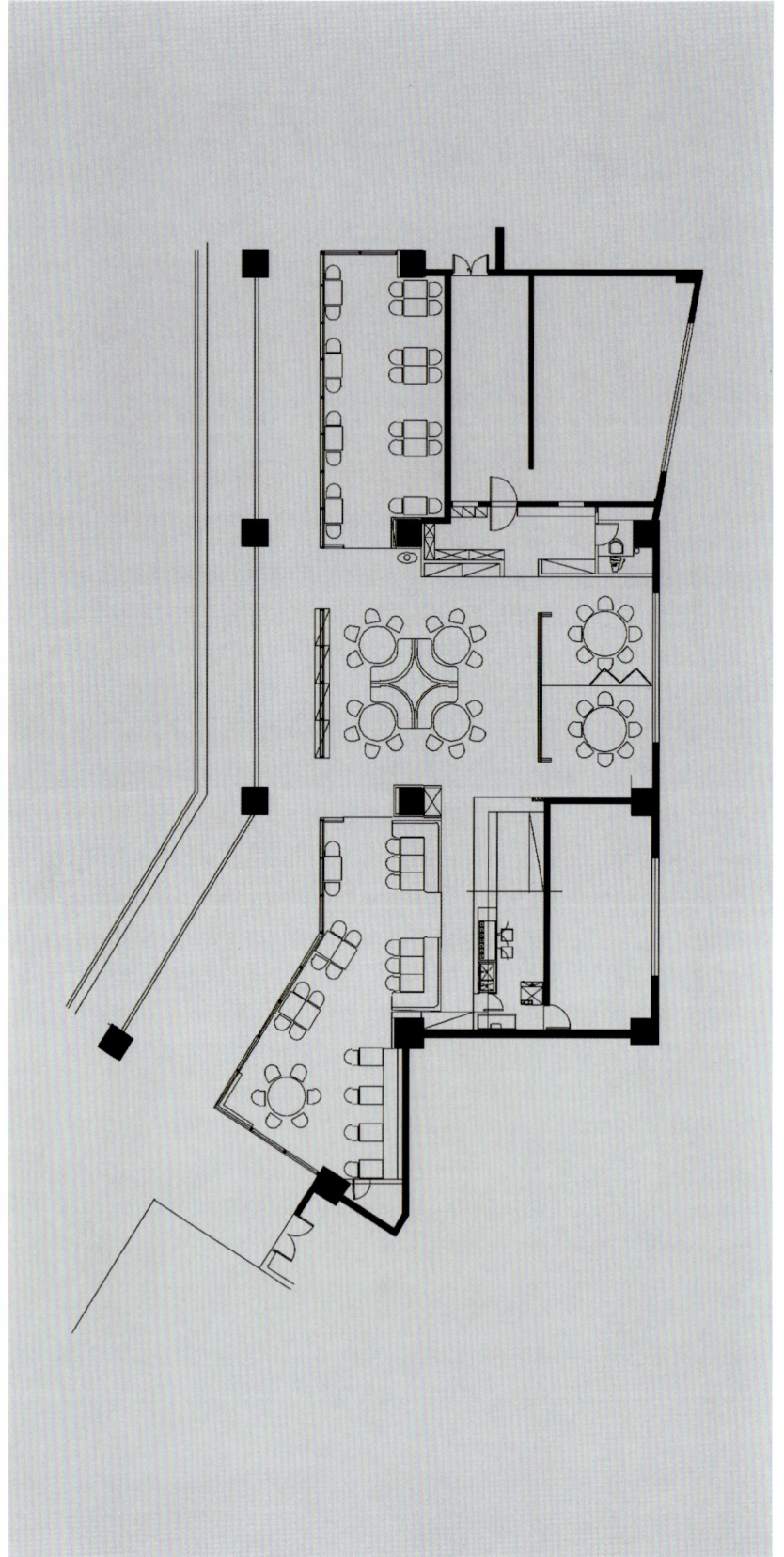

星星王子下凡

禾风韩式料理

　　本案是本地业主将韩式料理中国化的新尝试。以矩阵律动排列错落的木质格栅配以梦幻的色感，而非韩国的大正红色。洋红色天花板与光线交织的明暗感，减弱了空间高度不高的缺点，并利用色差减低压迫感。弧形在本案有一个非常有趣的展现：入口的两个半弧形墙，吻合着另一个弧形隧道座位区，墙面材质同样以石板立体呈现。灯光在此也有一次淋漓尽致的发挥。蓝色的LED灯与镜面包覆的柱子是全场最令人摒息的梦幻感，犹如星星王子所衷爱的遥远天堂。而灰色的椅背则含蓄大方的充当背景色。另外，全场地板高低垫高除了地面排烟技术面的考虑，MAY发挥客层入口，用餐顺畅的动线，吻合了使用及梦幻的DESIGN。

The present project is a new trial to localize the Korean cuisine in China by a local proprietor. The wood grates have been well arranged in the rhyme of rectangles with the coordination of the dreamy sense of colors instead of the genuine red color used in Korea. The ceiling in carmine color and the sense of brightness and darkness resulting from the mingling of lights have weakened the drawback of insufficient height of the space; and the oppressiveness has also been reduced with the use of the different shades in colors. The arc shape has a very interesting exhibition in the present project with two half-arc-shaped walls at the entrance coinciding with another arc-shaped seat area of the tunnel type. The material texture for the wall surfaces similarly employs the stone plates for the cubic demonstration. Here the lamp lights also have made a pleasurable expression. The blue LED lamps and the columns wrapped with the mirror surfaces have produced a dreamy sense for the people to hold their breaths just like the remote heaven loved by the star prince. The grey chair backs play the role of the background color in an obscure yet graceful way. Moreover, the floor boards in the entire area have been raised, for one thing, it is out of the consideration technically for the emission of smoking gas; for another, MAY has offered a full play to the fine-quality entrance for the customers and used the unobstructed moving line to coincide with the requirement on usage and the dreamy DESIGN.

主　设　计：张秀贞

参与人员：关中杰、赵立

摄　影　师：周宇贤

主要材料：冰花玻璃、木隔栅、LED 灯串

坐落地点：上海古方路南方休闲广场

面　　　积：430m²

完工时间：2006.12.24

协力厂商：米高装饰设计工程

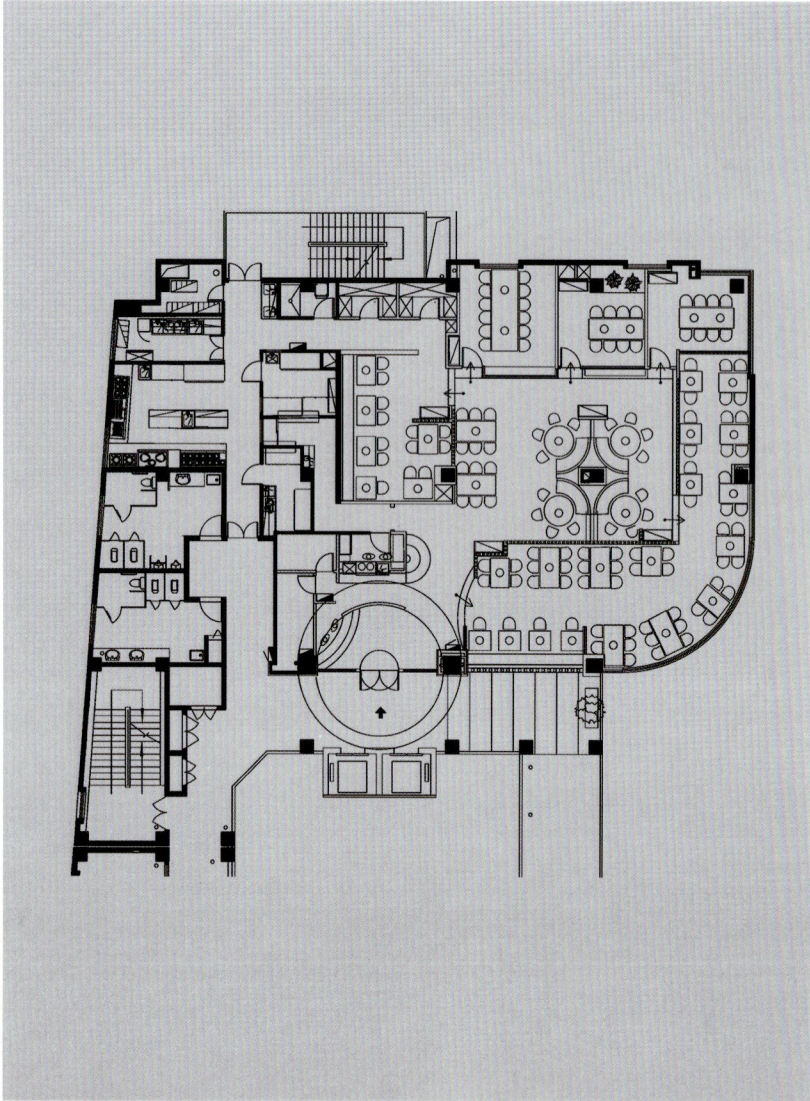

作品一览
Preface

◆ **1986–1990年台湾**

(设计)SOGO百货——台北市、中友百货——台中市、大统伊势丹新竹分馆改装工程——新竹店、大统伊势丹百货女装部改装工程 – 高雄市

(设计、工程管理)中兴三温暖SPA馆、美菲办公室、吴东升住宅、徐荷芳住宅、日盛证券工程、莲园（敦南）餐厅、聚利办公室

◆ **1991–1995台湾**

(设计、工程管理、工程施工)嘉琦服饰公司——北、中、南店、王秉文住宅、嘉琦服饰门市工程、Louis Freard——全台女装专卖店

◆ **1996–2000台湾**

(设计)板桥邱公馆

(设计、工程管理)无极天台庙建庙工程

(设计、工程管理、工程施工)伊凡特广告公司、赞泰周公馆、美伦办公室、韦耀华公馆、禾略广告公司、稳贸先进科技办公室

◆ **2001–2004年台湾**

(设计)桃园文山山庄许公馆

(设计、工程管理、工程施工)紫藤楼高公馆、开将高科技公司办公室 、南港区陈公馆

◆ **上海市**

(设计)食全食美集团集集小镇——金山店

(设计、工程管理)王品集团台塑牛排仙霞店、王品集团陶板屋日式料理

(设计、工程管理、工程施工)力怡皓远东国际办公室、东森集团总部办公室、睿晶服装工作室、世贸滨江张公馆、新华路王公馆、香谢丽舍潘公馆、张江汤臣高尔夫别墅刘公馆、东方曼哈顿陈公馆、阮公馆酒店式公寓、华光花园罗公馆、中芯A3别墅王公馆、曼都集团曼都美发——华山店、曼都集团曼都佳人——黄陂北路店、曼都集团曼都美发——古北店、曼都集团总部办公室、曼都集团曼都佳人——乌鲁木齐店、王品集团台塑牛排——长宁区仙霞店二期修正工程、王品集团台塑牛排张杨店设计修正——浦东新区

◆ **北京市**

(设计、工程管理、工程施工)王品集团台塑牛排——朝阳区国贸店

◆**江苏省启东市**

(设计、工程管理、工程施工)鼎尊企业太阳岛餐厅

◆**江苏省常熟市**

(设计)食全食美集团集集小镇

◆**2005上海市**

(设计、工程管理、工程施工)山水别墅凌公馆、王品集团西堤牛排——徐汇区天钥桥店、王品集团西堤牛排——黄浦区新世界店、王品集团西堤牛排——黄浦区淮海店、王品集团台塑牛排——卢湾区长乐店、豆捞坊澳门火锅——静安寺店、豆捞坊澳门火锅——浦东新区张杨店、泰平天国——泰式料理新世界店、竺石美妍SPA馆

◆**北京市**

(设计、工程管理、工程施工)王品集团台塑牛排——朝阳区西单店、悦堂火锅时代

◆**江苏省苏州市**

(设计、工程管理)王品集团台塑牛排——河滨店

◆**福建省厦门市**

(设计、工程管理、工程施工)叶公馆

◆**广东省深圳市**

(设计、工程管理、工程施工)王品集团台塑牛排——福田区深南店

◆**台湾**

(设计)钟妇产科——台北店

◆**2006上海市**

(设计、工程管理、工程施工)丽都新贵苏公馆、品哲配线办公室、EK咖啡——长风店、亚泰店、澜料理、豆捞坊澳门火锅——龙之梦店、禾风烧烤、王品集团丰滑火锅——百联店

◆**北京市**

(设计、工程管理、工程施工)王品集团丰滑火锅——崇光店

◆**江苏省苏州市**

(设计、工程管理、工程施工)王品集团西堤牛排——苏州狮山店、王品集团丰滑火锅——苏州狮山店、冉启集团789新概念料理餐厅——河滨店

◆**广东省深圳市**

(设计、工程管理)十大书坊

◆**2007上海市**

(设计、工程管理、工程施工)赤坂亭日式烧烤、赤坂亭日式烧烤——新天地店（筹划中）、豆捞坊澳门火锅——五角场（筹划中）、卢公馆（筹划中）、港味小厨——浦东新区正大广场店（筹划中）、红子鸡——澳门路店（筹划中）、申粤轩——丁香花园店（筹划中）

人们的思想使得设计的意境变得多么无限宽广!!!我们积极、乐观运作台湾实务经验与创作智慧。

1990年，一群建筑专业同学成立了May space for design studio，在生活与工作之间开始，体验空间设计的魔力，"maybe、maybe"的遥想空间漫流至今，所有自信心来自于自我坚持及周边人、事、物无限的祝福。

空间设计有许多共创语言，我们也将这挑战设计层次课程与业主分享，历练出新作品故事及装饰符号，让空间能透析说谈出归属于自己的餐饮文化。为了更完美演出实务，设计与工程间管理协调性是绝对重要，2003年至2006年间，我们得到客户的肯定，完成工程设计与工程管理的推荐书。总之，唯有尊重自我角色才能拥有未来设计立场的被尊重，也许这就是设计领域中所谓的成就感吧！

现阶段我们的团队——关中杰、张秀贞及全体同仁，依虚怀若谷的心态继续游学于大中国上海、北京、深圳等地。选择最活化的餐饮市场拓印出一系列充满情感空间设计的作品。也期待各方给May space更多支持与鼓励。

How infinitely wide the people's thinking has made the artistic conception of design!!! We're operating with the practical experience and creative wisdom in Taiwan in an active and optimistic way.

In 1990, a group of schoolmates majoring in architecture founded the May Space for Design Studio and began to experience the charm of space design. The space for free thinking by the wording "maybe, maybe" has extended up to the present time; and all the self confidence originates from the self persistence and the limitless blessings from the people, things and materials by our side.

Space design has many languages of joint creation. We also shared the challenge of the design layer course with the proprietors, worked out new stories and decorative symbols of the works with hard labor and let the space analyze and tell out the catering culture belonging to itself. For the better perfection of the practical business, the coordination of the design with the engineering management is absolutely necessary. During the years from 2003 to 2006, we obtained affirmation from the customers and completed the letter of recommendation for the engineering design and engineering construction. In a word, only by respecting the self role can we possess the future stand of design to be respected by the other people. Maybe this is the sense of achievement so-called in the sphere of space design!

At the present stage, our team (Guan Zhongjie, Zhang Xiuzhen and all the colleagues) are continuing our studying Shanghai, Beijing, Shenzhen and some other places of Grand China for selecting the most activated catering market for printing out a series of works of space design full of emotions and looking forward to more support and encouragement from various parties concerned to May Space.

图书在版编目 (CIP) 数据

餐饮空间设计 / 黄玉枝编 . — 沈阳 : 辽宁科学技术
出版社 , 2017.6
ISBN 978-7-5591-0192-1

Ⅰ . ①餐… Ⅱ . ①黄… Ⅲ . ①饮食业－服务建筑－
建筑设计 Ⅳ . ① TU247.3

中国版本图书馆 CIP 数据核字 (2017) 第 072627 号

出版发行：辽宁科学技术出版社
　　　　　（地址：沈阳市和平区十一纬路 25 号　邮编：110003）
印　刷　者：辽宁新华印务有限公司
经　销　者：各地新华书店
幅面尺寸：250mm × 250mm
印　　张：22⅓
插　　页：4
字　　数：80 千字
出版时间：2017 年 6 月第 1 版
印刷时间：2017 年 6 月第 1 次印刷
责任编辑：宋丹丹 李亮亮
封面设计：李　莹
版式设计：李　莹
责任校对：周　文

书　　号：ISBN 978-7-5591-0192-1
定　　价：228.00 元

编辑电话：024-23280367
邮购热线：024-23284502
E-mail: 1207014086@qq.com
http://www.lnkj.com.cn